Robert Geroch

General Relativity

1972 Lecture Notes

MINKOWSKI
Institute Press

Robert Geroch
Enrico Fermi Institute
University of Chicago

Cover: Lecture notes are often written in similar environments

ISBN: 978-0-9879871-7-4 (softcover)
ISBN: 978-0-9879871-8-1 (ebook)

Minkowski Institute Press
Montreal, Quebec, Canada
http://minkowskiinstitute.org/mip/

For information on all Minkowski Institute Press publications visit our
website at http://minkowskiinstitute.org/mip/books/

Preface

This publication of Robert Geroch's course notes on general relativity is the first volume in the new *Lecture Notes Series* of the *Minkowski Institute Press*. The idea of this series is to extend the life in space and time of valuable course notes in order that they continue to serve their noble purpose by bringing enlightenment to the present and future generations.

Geroch's lecture notes on general relativity are unique in three main respects. First, the physics of general relativity and the mathematics, which describes it, are masterfully intertwined in such a way that both reinforce each other to facilitate the understanding of the most abstract and subtle issues. Second, the physical phenomena are first properly explained in terms of spacetime and then it is shown how they can be "decomposed" into familiar quantities, expressed in terms of space and time, which are measured by an observer. Third, Geroch's successful pedagogical approach to teaching theoretical physics through visualization of even the most abstract concepts is fully applied in his lectures on general relativity by the use of around a hundred figures.

Although the book contains lecture notes written in 1972, it is (and will remain) an excellent introduction to general relativity, which covers its physical foundations, its mathematical formalism, the classical tests of its predictions, its application to cosmology, a number of specific and important issues (such as the initial value formulation of general relativity, signal propagation, time orientation, causality violation, singularity theorems, conformal transformations, and asymptotic structure of spacetime), and the early approaches to quantization of the gravitational field.

Montreal, February 2013   *Vesselin Petkov*
                                    Minkowski Institute Press

Contents

1. Introduction

The general theory of relativity is the most attractive classical (i.e., non-quantum) theory of the gravitational field available today. It represents a "generalization" of Newtonian gravitation which is consistent with the principles of special relativity.

Recall Newtonian gravitation. The gravitational potential is determined from the distribution of matter by the equation

$$\nabla^2 \varphi = 4\pi G \rho \tag{1}$$

where G is the gravitational constant and ρ is the density of matter. This potential then produces a force, $F = -\nabla\varphi$, on a particle of mass m. Thus, a distribution of matter produces a potential φ which, in turn, affects particles: in this way bodies interact gravitationally. What is wrong with this theory? The problem is that the distribution of matter at any one instant of time determines the potential at that time. This action is instantaneous. In other words, a particular "time" is singled out. Such a situation is simply not permitted by the principles of special relativity.

Thus, the phenomenon of gravitation certainly exists, and special relativity is about as well-established as a physical theory can be. One is naturally led to search for a relativistic theory of gravitation.

One can think of special relativity as the "limit" of general relativity when gravitation is unimportant, and of Newtonian gravitation as the limit when speeds are much less than that of light (so the Newtonian picture of space and time is a good approximation to the special relativistic picture). (Thus, the statement one occasionally hears that general relativity must replace special relativity when "accelerated frames" are considered is false. Special relativity must be abandoned when gravitation is important.)

It turns out that the general theory of relativity (i.e., this incorporation of gravitation into special relativity) touches on the structure of space and time. Why should this be true for gravitation (and not, say, for electromagnetism)? It is because gravitational phenomenon has about it an "all-inclusiveness" which is apparently not possessed by other forces in Nature (e.g., electromagnetic). Thus, particles can be positively charged, negatively charged, or uncharged: particles can react to electromagnetic fields in different ways. But all particles seem to react to gravitation in the same way. (Quantum theory is a good example of a physical theory with a similar "all-inclusiveness." The principles of quantum mechanics are not to be applied just to certain special systems. Quantum theory makes a much stronger claim. All physical phenomena are to be governed by its principles.)

This feature of gravitation is made more explicit by considering the famous elevator. People inside an elevator falling freely in a uniform gravitational field cannot distinguish their situation from that of an elevator at rest in no gravitational field. The uniform gravitational field affects everything – so by no physical experiment inside the elevator can these two situations be distinguished. (Like many important physical principles, this, the equivalence principle, becomes a tautology in the final theory. In general relativity, "uniform gravitational field" is a long way of saying "no gravitational field." That is, the existence of the former is simply not recognized. More on this later.)

The remarks above at least make it reasonable that gravitational phenomena might occupy a special place in physics. They do not, however, make it clear why this "special-ness" should manifest itself in statements about space and time. That this should be the case is suggested by the remarks below.

In order to do Euclidean geometry, one needs straight lines. I draw a line in the room. How do you determine whether or not it is straight? One might think of putting the edge of a meter stick against it. But, of course, that doesn't work unless one has a way to tell which meter sticks have straight edges. One might stretch a string between two points on the line, and see whether the string lies along the line. But a gravitational field will bend the string in a way over which we do not exert control. (The complete explanation here is more subtle. We shall return to this.) Finally, one might "sight" along the line to see if it is straight. Here, the implicit assumption is that light travels along straight lines. But it is known that a gravitational field affects the path of a light ray, for light from distant stars is observed to be bent of passing near the surface of the sun. Similar attempts fail for similar reasons. In short, since gravitation affects all physical phenomena (all-inclusiveness), and since such phenomena must be used, in particular, to make the measurements which determine the geometry of space (and time), it is perhaps not surprising that gravitation should be tied in with the structure of space and time.

2. Events; Space-Time

Recall the Newtonian picture of physics. One has some physical system he wishes to describe. At any one instant of time, that system is in one of a certain collection of possible states. As time goes on, the system evolves, i.e., it passes through a sequence of states. This general approach to describing a physical system is so natural that it is not always mentioned explicitly. We mention it because the description of things in relativistic physics is quite different. Time loses its special role. Instead, one adopts a description in which the past, present, and future of the system is all incorporated into one picture. This idea is reflected in the following.

A fundamental notion of general relativity is that of an *event*. An event is an occurrence (in the physical world) having no extension in either space or time. For example, the snapping of one's fingers, or the exposition of a firecracker, would represent an event.

The collection of all possible events – those which have happened, are happening, or ever will happen – will be denoted by M, and called *space-time*. This space-time is the arena in which all the physics takes place. An event is sometimes called a *point* of M.

The space-time of M of general relativity has more structure than just that of a set. We introduce part of this structure now. Suppose we consider a certain region of M (e.g., the collection of events which occur within a given room within of given interval of time). We wish to label points of this region by numbers. This could be done, for example, as follows. Let four pilots fly airplanes around the room. Each pilot carries with him a clock. (These clocks do not have to run at a uniform rate, or be synchronized.) Let an event which takes place in our region be the explosion of a firecracker. Each pilot records the time (as read from his clock) when he sees the flash. Thus, with this event, we associate the four numbers obtained by the pilots. These four numbers are called the *coordinates* of the event. Evidently, no two distinct events have exactly the same coordinates, and any set of coordinates (in the appropriate intervals) defines an event.

Evidently, there are many different systems of coordinates. The pilots can fly around in arbitrary ways, using clocks that run at arbitrary rates. Furthermore, there are other physical ways of obtaining systems of coordinates. One could, for example, fill the room, from floor to ceiling, with people during the time interval. Each person could be assigned three numbers which give his location in the room, and each person could carry a watch. Then, if an event takes place in the region of space-time, the person located at the occurrence of that event could record his watch's reading when the event takes place, and the three numbers which give his location. In that way, again, events are labeled by four real numbers.

Now suppose we select a region of space-time, and have two systems of coordinates, x^1, x^2, x^3, x^4 and y^1, y^2, y^3, y^4, in this region. That is to say, each event in our region is labeled by four numbers in two different ways. Hence (since values for the y-coordinates determine an event in the region, which, in turn, determines values for the x-coordinates), we obtain four functions of four variables, $x^1(y^1, y^2, y^3, y^4)$, $x^2(y^1, y^2, y^3, y^4)$, $x^3(y^1, y^2, y^3, y^4)$, $x^4(y^1, y^2, y^3, y^4)$. We assume that these functions are smooth (c^∞). (This is reasonable physically if the coordinate systems are set up with sufficient care. Thus, the pilots would have to move the controls of their airplanes smoothly, and their clocks would run smoothly, etc.) More generally, if the x's are an admissible system of coordinates, and if the y's are smooth functions of the x's, with the inverse functions (the x's as functions of the y's) also smooth, we permit the y's as a system of coordinates.

Thus, on the space-time M we have a large number of admissible coordinate systems, any two smoothly related to each other, and with any system smoothly related to the admissible ones also admissible. But this is what the mathematical object called a four-dimensional manifold is. Thus, *the set M of events has the structure of a four-dimensional manifold*. (Such statements, which relate the physics to its mathematical description, are never proven. All one can do is make the mathematical description appear appropriate. Nor is it claimed that one can see, by abstract reasoning, that such a description will be useful. The foundation of a theory can be motivated, but its testing can only occur after the structure of the theory has been set down in detail.)

We wish to adopt the following point of view: every statement which is made about physical phenomena is to be expressed in terms of the space-time manifold M. This is somewhat different point of view from that of Newtonian physics. The remarks bellow are intended to make the new attitude seen more natural.

How would one describe a particle in terms of space-time? Consider the set of all events which occur at the particle. That is, we consider events (snapping of fingers) such that, at the instant the fingers are snapped, the particle is right there. Evidently, this set of events describes a curve in space-time. This curve, called the *world-line* of the particle, describes completely the entire history (and future) of the particle.

Similarly, a string is described by the collection of all events which occur on the string – a two-dimensional surface in space-time. A bedsheet defines a three-dimensional surface in space-time. Two particles which 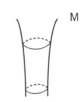 collide have the property that their world-lines intersect. The point of intersection is the event of the collision of the particles.

Note that there is no dynamics in space-time: nothing ever happens there. Space-time is an unchanging, once-and-for-all picture encompassing past, present, and future.

A star would be represented by a four-dimensional region of space-time. It would be tubular shaped, with (qualitatively) time going up the tube. If the star suddenly exploded, then the tube would "get wider" beyond a certain point. Collapse would be represented by a contracting of the tube.

There is, as we shall see later, no natural notion of "simultaneity" between events in space-time. But everyone near the Earth agrees when it is 1 : 00 Greenwich mean time. Suppose we fill the Earth (surface, interior, atmosphere) with people, all instructed to snap their fingers (mark the event) at 1 : 00 Greenwich mean time. These events would form a three-dimensional surface. If fingers were again snapped at 1 : 01, 1 : 02, etc,. we would obtain a family of such surfaces, filling the space-time. We can consider the first surface as representing "all of space at 1 : 00," the next as "all of space of 1 : 01", etc. We have now recovered simultaneity: two events are simultaneous if they occur on the same surface. We have also recovered dynamics: the situation at 1 : 00 is described by the space-time on the 1 : 00 surface, the situation at 1 : 01 by the 1 : 01 surface, etc. Things "change with time", for, as one moves from one surface to the next, one enters different regions of space-time, and so what is going on there might be different. A particle (i.e., a curve in space-time) would meet the successive surfaces at different points, and thus would be described as "moving in space as time evolves".

In short, we recover the Newtonian picture of physics by slicing the space-time with a family of three-dimensional surfaces.
But, of course, there are many
different ways to carry out such
a slicing. There is no natural
slicing, hence, no natural simultaneity, etc. All that is "natural"
is the entire, four-dimensional
space-time manifold. (There is

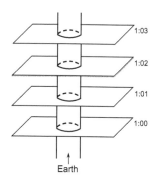

nothing philosophical about this. It is just a different – and a rather
attractive – way of describing what is going on.)

To summarize, an event is an
occurrence having extension in
neither space nor time. All the
possible events (that have happened, are happening, or will
happen) are collected together
into space-time M. This M
has the structure of a four-
dimensional manifold. One
adopts the point of view that ev-

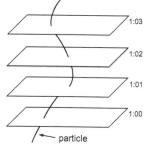

erything is to be described in terms of structure on this manifold M.

3. The Light-Cone Structure

The next piece of structure on our space-time manifold M arises from the observation (in the real world) that no particles have been observed to move faster than light. thus the motion of light rays is of special significance.

The word "faster" refers to "speed", which refers to "distance traveled per unit time". Hence, the notion "faster" violates the spirit of space-time physics. We reformulate the observation in the first paragraph as follows. Suppose the explosion of a fire-cracker marks an event, so an expanding spherical pulse of light is sent out from it. Then any particle whose world-line passes through this event cannot get outside of the expanding sphere of light. (This is the statement that the particle's speed doesn't exceed that of light.) Let's try to reformulate this as a statement in space-time. Let the event (fire-cracker) be p. Then the light emerging from this event would form a cone with vertex p. (Slice the space-time with a family of three dimensional surfaces, and you will, indeed, see a "sphere of light expanding from the original event".) (In the figure, one dimension is, of necessity, suppressed.) Now we consider a particle (i.e., a curve) passing through p. The observation (about speeds not exceeding that of light) now reduces to the assertion that the world-line of the particles is inside the cone with vertex p.

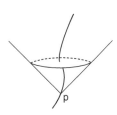

Thus, for each event of space-time there is a cone with vertex that event. This cone is the events which occur just as light from the original event passes by. The world-lines of particles which pass through the original event lie inside the cone. The cone is called the (future) *light cone*.

Now let p and q be two nearby events. (We need "nearby" because some notions, logically different, coincide for nearby events but do not coincide for more widely separated events. Of course, "nearby" means nearby in space-time.) If q lies inside the future light cone of p, we say that q is *timelike related* to p and to the *future* of p. If p lies inside the future light cone of q, we say that q is timelike related to p and to the *past* of p. If q lies on the future light cone of p (resp., p lies on the future light cone of q), we say that q is *null related* to p, and to the future (resp., past) of p. Finally, if neither event is either on or inside the future light cone of the other, we say the events are *spacelike related*. Evidently, a particle can meet first p and then q if and only if these events are timelike related, and p is to the past of q. A light ray can meet both events if and only if they are null related. Spacelike related events cannot be both met by any particle or light ray. (Note: A "light ray" corresponds to a flashlight which is on only momentarily, sending out a pulse of light. The light from a flashlight which is on continuously defines a two-dimensional surface in space-time.)

It should now be clear that the three-dimensional surfaces introduced earlier (to obtain simultaneity) had the property that the future light cone from any event on any surface does not elsewhere intersect that surface. (Physically: light emitted at $1:00$ cannot get to any other point by $1:00$.)

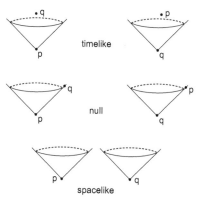

In special relativity, one can arrange matters, in an appropriate coordinate system, so that all the light cones look like ordinary cones, all with the same opening angle, and all parallel to each other. There is no reason, a-priori, to make such an assumption. In general relativity, no such assumption is made. Dropping it amounts to admitting curvature to space-time.

To summarize, one constructs a future light cone at each event using a light pulse emitted from that event. Particle world lines which pass through the event remain inside the light cone. The light cone leads to a distinction between timelike, spacelike, and null re-

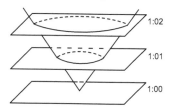

lated events, and to a distinction between timelike, spacelike, and null related events, and to a distinction between past and future related events. (The past, future distinction is not available for spacelike related events.)

4. Clock Rates

We shall shortly obtain a geometry (more precisely, a metric tensor field) on the space-time manifold M. This geometry is determined physically by measurements of time intervals between certain pairs of events. Finally, the possibility of making meaningful measurements of elapsed times rests on a certain assumption. We now introduce this assumption.

There exists clocks. We suppose that it is feasible to build clocks whose workings are not significantly affected by external influences (e.g., electromagnetic fields, accelerations, etc.). (This is a rather vague statement, whose meaning tends to change as our understanding of physics changes.) Now suppose we build two such clocks. Suppose, furthermore, that it is found that, when the clocks are held side by side (in space-time language, when the world-lines of the clocks are made to practically coincide over a certain stretch) their tick-

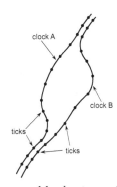

ing rates coincide. (For example, this would, presumably, be true of two Cesium clocks.) Now the clocks are separated, and, for a while, traverse different paths in space-time. Finally, at some later time, the clocks are again held side by side. We assume that, *if the clocks initially ticked at the same rate, then when they are again brought side by side, they again tick at the same rate*. Note that we do not assume that the total elapsed times measured by the clock while they are separated are the same. In fact, general relativity predicts that this will not in general be the case. The assumption, essentially, is only that the ticking rate of an "isolated" clock along a certain stretch of its world-line depends only on the world-line in that region, not on what the world-line was before the region of interest. In other words, the ticking rate of a clock does not depend on what has happened to it in the past.

Why should one believe such an assumption? It would imply, for example, that two atomic clocks, one assembled in Chicago and the other in Bombay will, when they are brought together, tick at the same rate. Physically, one might expect this to be the case.

An "experimental test" of this assumption has been suggested by Wheeler. An electron has a mass, and hence a Compton wavelength, and hence a characteristic time (the light-travel time across a Compton wavelength). Hence, an electron is essentially a "clock". The electrons in an atom inside the Earth have, presumably, traversed different world-lines in space-time to get where they are. A "comparison of the ticking rate" is provided by the Pauli exclusion principle. It is because of this principle (i.e., because all electrons are identical) that after the first

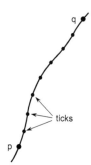

few electrons in an atom have occupied the lowest states, the remaining electrons must occupy higher states. Now, a typical electron in an atom in the Earth has gone around the nucleus 10^{30} times during the age of the Earth. Hence, if mass differences between different electrons were more than about one part in 10^{30}, all the electrons would by now have dropped to the lowest level (because they would have "discovered" their mass differences, i.e., discovered that they do not have to obey the exclusion principle). But this collapse of atoms in the Earth has not taken place. Hence, our assumption has been "tested" to one part in 10^{30}.

Now suppose we build one clock, and call the elapsed time between its ticks the second. Then any other clock can, by being brought side by side with this one, be recalibrated so that it ticks in seconds too. Once the second clock has been so calibrate, it can be carried around anywhere in space-time. We calibrate a large number of clocks in this way. Then, by our assumption, any two will, when brought side by side, tick at the same rate. In other words,if we have a world-line of a particle, and two events p and q on this world-line, then we can count the number of ticks (number of seconds) between the events p and q using a calibrated clock which traverses the world-line between p and q. Our assumption ensures that this number (of seconds) is independent of which calibrated clock we use. (Although it will in general depend on what the world-line is between p and q.)

Thus, *given a world-line between events p and q, it is meaningful to speak of the elapsed proper time (in seconds) along that world-line between p and q.*

5. The Interval

We can now speak of the elapsed proper time between two events along a world-line joining those events. This leads to the notion of the interval between two events.

Let p and q be two nearby events in space-time. (More precisely, we require that the displacement of q from p be infinitesimal. That is, what follows is good only to first order in the displacement.) Arrange a world line of a particle to pass through event p. Now find a light ray which first intersects this world-line, and then event q. Find a second light ray which first meets q, and then intersects the world-line. Denote by u_1 and u_2 the two events at which 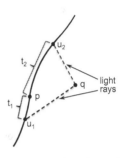 these light rays intersect the world-line. Finally, let t_1 be the elapsed proper time along this world line between u_1 and p, and let t_2 be the elapsed proper time between p and u_2. (Thus, t_1 and t_2 are just a certain number of seconds.) We define the *squared interval* betwe

$$SI(p,q) = t_1 t_2 \tag{2}$$

We next require a further assumption: that the quantity $SI(p,q)$ depends only on the (nearby) events p and q, and not on the other details of the construction (whether the original world-line is passed through p or q; the choice of that world-line).

What are the grounds for this assumption? Let us see what the construction above yields in special relativity. Our space-time

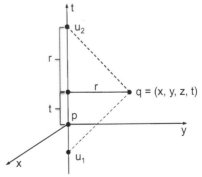

is Minkowski space, with the usual x, y, z, t coordinates. Let p be the event with coordinates $(0,0,0,0)$, and q the event with coordinates (x, y, z, t). Let the world-line be the t-axis (the straight line $x = y = z = 0$). Then the parameter t along this world-line measures proper time (between events on this world line). Light rays are straight lines making a $45°$ angle to the vertical (i.e., to the t-axis). That is, the light cones are all open $45°$ from the vertical. In this case, the event u_1 has coordinates $(0,0,0,t-r)$, and u_2 coordinates $(0,0,0,t+r)$, where $r^2 = x^2 + y^2 + z^2$. Hence, $t_1 = r - t$, and $t_2 = r + t$. Thus, $SI(p, q) = t_1 t_2 = r^2 - t^2 = x^2 + y^2 + z^2 - t^2$. This quantity will be recognized as the usual squared interval between points in Minkowski space in special relativity. Clearly, the construction above yields the correct squared interval for any two events, and any straight world-line in special relativity.

Of course, even in special relativity, the quantity $SI(p, q)$ will not be independent of the world-line if non-straight world-lines are admitted. It is clear intuitively, however, and easily checked, that independence of the world-line does hold in special relativity to first order in the displacement of q from p.

Thus, our assumption amounts, essentially, to the assumption that, *locally, space-time has the same structure as the space-time of special relativity*. More precisely, space-time is that of special relativity to first order in displacement. (Still more precisely, and in the language of general relativity, the displacement of p from q must be small compared with the radius of curvature of space-time.)

In the discussion above, q was spacelike related to p; and we had $SI(p, q)$ positive. Suppose that q were null related to p. Then $u_1 = p$ (if q is to the future of p), or $u_2 = p$ (if q is to the past of p). This is clear because "null related" means precisely that there is a light ray joining p and q. Hence, when q is null related to p, either $t_1 = 0$ or $t_2 = 0$ (the cases $u_1 = p$ and $u_2 = p$, respectively). In either case, $SI(p, q) = t_1 t_2 = 0$.

Suppose, finally, that q is timelike related to p. Then, if q is to the future of p, u_1 and u_2 are to the future of p, so t_1 is negative and t_2 positive. If q is to the past of p, then u_1 and u_2 are both to the past of p, so t_1 is positive and t_2 negative. Thus, in either case, $SI(p, q)$ is negative.

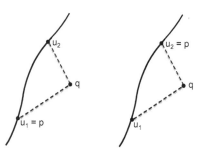

To summarize, $SI(p, q) = SI(q, p)$ *is positive if p and q are spacelike related, zero if p and q are null related, and negative if p and q*

are timelike related.

Now suppose that p and q are timelike related (say with q to the future of p). Choose the world-line to pass through p and q. Then $u_1 = u_2 = q$ (since q is already on the world=line, the "light rays" from q to the world-line are very short). Hence, $-t_1 = t_2 =$ proper elapsed time between p and q along this world-line. (To first order, this quantity is independent of world-

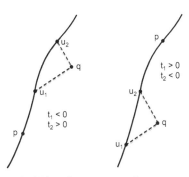

line). Hence, *for p and q timelike related, $SI(p, q)$ is minus the square of the proper elapsed time between p and q* (along any world-line joining the events).

Similarly, let p and q be spacelike related, and choose the world-line so $t_1 = t_2$. An observer on this world-line would interpret the situation as follows. He is bouncing a light ray off q, and measuring the round-trip light travel time, $t = t_1 + t_2$. He would say that the "spatial distance" of q from p is $\frac{1}{2}t$ (where his unit of distance is the light-second, the distance light goes in a second). That is, *for p and q spacelike related, $SI(p, q)$ has the intuitive interpretation as the square of "distance" (measured in light-seconds) of q from p.*

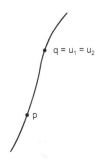

In this sense, then $SI(p, q)$ is interpreted as representing geometrical information about spacetime.

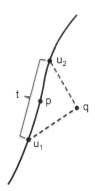

6. The Metric

Imagine a very large book in which there is listed all possible pairs, p, q, of nearby events, and, for each pair, the number $SI(p, q)$. This book contains all the information about the geometry of space-time. One could in principle treat the geometry of space-time by constantly dragging out this book, but obviously, this would not be very convenient. One is thus let to attempt to describe the geometrical information in SI in a more convenient form.

Let p be an event of space-time, i.e., let p be a point of the manifold M. Recall that a nearby event, q, to p, is described by a contravariant vector, ξ^a, at p. We now introduce the assumption: there is a tensor $g_{ab} = g_{ba}$ at p such that

$$SI(p, q) = g_{ab}\xi^a\xi^b \qquad (3)$$

for all "infinitesimally nearby" q. This assumption can be considered as a further extension of the idea that "locally, space-time has the structure of space-time in special relativity". In special relativity, expression (3) holds, where g_{ab} is the usual metric $(dx^2 + dy^2 + dz^2 - dt^2)$ of Minkowski space. We therefore assume that it holds at each point of space-time.

The above is repeated at each point of space-time. In this way, we obtain a symmetric tensor field g_{ab} on the manifold M. Since ξ^a connects infinitesimally nearby points which are null related precisely when $g_{ab}\xi^a\xi^b = 0$, and since the points null related to a point p are supposed to form a cone with vertex p, it is clear that this g_{ab} must have signature $(-, +, +, +)$. (Or, to state matters another way, we have special relativity locally.) We assume that g_{ab} is a smooth tensor field. (As a rule of thumb, whenever smoothness is meaningful for a physical quantity, there is no loss, physically, in assuming smoothness. Perhaps that's a reason why differential geometry – the study of smoothness – is so useful in physics.) From the statement of its signature, g_{ab} has an inverse, g^{ab}. Thus, g_{ab} is a metric tensor field of M.

To summarize, *the information in the $SI(p,q)$'s i.e., the geometry of space-time, is completely described by a metric tensor field g_{ab} on M of signature $(-,+,+,+)$*.

For obvious reasons, a vector ξ^a at a point is called *timelike, null,* or *spacelike* according as $g_{ab}\xi^a\xi^b$ is negative, zero, or positive, respectively.

We now have a four-dimensional manifold M with metric tensor field g_{ab} of signature $(-,+,+,+)$. All the technology of differential geometry can now be carried over to M. In particular, we have the notion of (indexed) tensor fields on M. Induces of such fields are raised and lowered with g_{ab} and its inverse. We have the (unique) derivative operator ∇_a on such tensor fields satisfying $\nabla_a g_{bc} = 0$. We have a Riemann tensor R_{abcd} satisfying the Bianchi identities, and having the algebraic symmetries $R_{abcd} = R_{[cd][ab]}, R_{[abc]d} = 0$. We construct from the Riemann tensor the Ricci tensor and scalar curvature.

The basic philosophy is that all physics is to be described in terms of space-time manifold M. We can now state this philosophy a bit more precisely: physics is to be described in terms of tensor fields on M. Among these fields, easily the most important is the metric, g_{ab}. There will soon be others.

We are now in a position to make more precise the distinction between special relativity and general relativity. *The framework we have set up above is called special relativity when the Riemann tensor vanishes* (everywhere), *and general relativity otherwise.* (It is easily checked that the vanishing of the Riemann tensor is precisely the condition for the existence of a coordinate system in which the components of the metric take the usual special relativity form, diag $(-1,+1,+1,+1)$.) Suppose that one views Minkowski space (the space-time of special relativity) from a "uniformly accelerated frame", e.g., an accelerated elevator. No matter how the situation is "viewed", the Riemann tensor vanishes. In other words, special relativity is applicable.

It is clear, from the way the metric was introduced, that it represents geometrical information about space-time. as we shall see shortly, the metric has a second interpretation: as the "potential of the gravitational field".

We conclude with an obvious remark. The metric is uniquely determined by certain physical measurements (of SI). Hence, any statement one makes about the metric automatically has physical significance. Of course, some statements (about g_{ab}) may be easier to interpret than others, but all are in principle interpretable. One can freely do mathematics, introducing physical interpretations where they are appropriate and interesting.

7. The Geometry of World-Lines

We are now in a position to describe more precisely the world-lines of particles and light rays.

Consider a curve in space-time, i.e., a smooth mapping $\gamma\colon I \to M$, where I is some open interval of the real line. then, for each number λ in I, the tangent vector to this curve, ξ^a, is a contravariant vector at the point $\gamma(\lambda)$ of I. This curve is said to be *timelike* if, for each value of λ, the tangent vector ξ^a is timelike; *null* if, for each λ, ξ^a is nonzero and null; and *spacelike* if, for each λ, ξ^a is spacelike. (Note: timelike and spacelike imply nonzero.)

The statement of Sect. 3 is now the following: *the world-line of a material particle is a timelike curve; the world-line of a light ray is a null curve.* (This whole business is a bit circular, as such arguments always are. We originally used the world-line structure to obtain the metric; now that structure is reexpressed in terms of the metric.)

Let us reparameterize our curve. Let $\lambda(\lambda')$ be a function of λ', so we replace the parameter λ by λ'. Then the corresponding tangent vectors, ξ^a and ξ'^a, are related by

$$\xi'^a = \xi^a \frac{d\lambda}{\lambda'} \tag{4}$$

If p and q are two events on this world line, then, immediately from (4),

$$\int_p^q [-\xi'^a \xi'_a]^{1/2} d\lambda' = \int_p^q [-\xi^a \xi_a]^{1/2} d\lambda \tag{5}$$

Thus, the integral (5) depends only on the world-line and on the events p and q, not on the parameterization of the world-line. This number is called the *length* of the (timelike) world-line between p and q.

There is a simple physical interpretation for the length of a timelike world-line. Recall that $(-\xi^a \xi_a)^{\frac{1}{2}}$ is the proper elapsed time be-

tween the events whose infinitesimal displacement is represented by
ξ^a. Hence, (5), the sum of these "infinitesimal elapsed times between
successive points on the world-line" is just the elapsed proper time
between p and q along this world-line.

Thus, we have that *the length of a timelike
curve between two events is precisely the elapsed
proper time* (in seconds) *between those events
along the curve.* (This statement could have served
as the physical definition of the metric.)

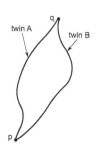

Suppose that two identical twins traverse
world-lines which meet at event p, and then again
at event q. In general, these two curves will not
have the same length between p and q. Thus, in
general, one twin will age more than the other be-
tween p and q. (This remark is sometimes called the "twin paradox".
Note that the fact that different curves between two points have different length doesn't require curvature. The difference in ageing is a
special-relativistic effect.)

It is clear from Eqn. (4) that, given a timelike curve, one can always
choose a parameter t for that curve such that its tangent vector is unit:
$\xi^a \xi_a = -1$. (Such a t is uniquely determined up to the addition of a
constant.) In this case, by (5), the proper time between two points on
the curve is just the difference between the t-values of those points.
One says that the curve is *parameterized by length,* or *parameterized
by proper time.* The corresponding (unit) tangent vector to the curve
is called the four-velocity (or just velocity) of the particle.

When we speak of an *observer,* we normally mean merely a timelike
curve. (You and I, after all, are merely timelike curves in space-time.)

Consider an observer, and consider two events,
p and q, with p on his world-line, and q infinitesimally displaced. Under what conditions would
our observer say that that the events p and q occurred "simultaneously". One criterion our observer might use is the following. He might send
out a light ray to intersect q, and receive a light
ray from q (just as if he were going to determine
$SI(p,q)$.) He might call the events "simultaneous"
if $t_1 = t_2$. (Think about that a moment. If you

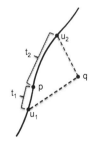

sent a light ray to the sun at 1:52, and it reached the sun just as a solar
flare began, and if you received a light ray from the flare at 2:08 (i.e.,
you saw the flare at 2:08), then you would say that the event "start of
the flare" and "when my watch read 2:00" were simultaneous.) Let's
express this condition $t_1 = t_2$ tensorially. Let χ^a be the vector repre-

senting the infinitesimal displacement between p and q, and let ξ^a be the four-velocity of the observer at p. Then $\lambda^a + t_1\xi^a$ is the displacement vector from u_1 to q, while $\lambda^a - t_2\xi^a$ is the displacement vector from u_2 to q. Since these displacements represent the paths of light rays, they must be null:

$$0 = (\lambda^a + t_1\xi^a)(\lambda_a + t_1\xi_a) = \lambda^a\lambda_a + 2t_1\lambda^a\xi_a - (t_1)^2$$
$$0 = (\lambda^a - t_2\xi^a)(\lambda_a - t_2\xi_a) = \lambda^a\lambda_a - 2t_2\lambda^a\xi_a - (t_2)^2 \tag{6}$$

These two equations can hold with $t_1 = t_2$ when and only when $\xi^a\lambda_a = 0$. In other words, *"pure spatial displacements" from p (according to this observer) are represented by vectors at p which are orthogonal to the four-vector of the observer.*

Again, let p be an event on our observer's world-line, and let q by a nearby event. Denote by η^a the vector representing this displacement. Our observer would say that q occurred a certain "spatial distance" from p, and a certain "time earlier (or later)" than p. He would obtain such a description as follows. He first decomposes η^a into the sum of a multiple of his four-velocity and a vector orthogonal to his four-velocity:

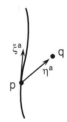

$$\eta^a = -\xi^a(\eta^m\xi_m) + (\eta^a + \xi^a(\eta^m\xi_m)) \tag{7}$$

The first term represents what he would call the "temporal displacement" between p and q. Hence, he would say that the "elapsed time" between these events is:

$$t = [-(\xi^a(\eta^m\xi_m))(\xi_a(\eta^n\xi_n))]^{\frac{1}{2}} = -\eta^m\xi_m \tag{8}$$

The second term in (7) represents what he would call the "spatial displacement" between p and q. (This term is orthogonal to his four-velocity, and so represents, for him, a spatial displacement.) Hence, he would say that the "spatial distance" between p and q is

$$r = [(\eta^a + \xi^a(\eta^m\xi_m))(\eta_a + \xi_a(\eta^m\xi_m))]^{\frac{1}{2}} = [\eta^a\eta_a + (\xi^a\eta_a)^2]^{\frac{1}{2}} \tag{9}$$

Note, from (8) and (9), that

$$r^2 - t^2 = \eta^a\eta_a$$

i.e., the squared interval between p and q. A different observer (i.e., a different world-line through p) would assign different "spatial and temporal displacement" to p and q. But all would agree on the value of $r^2 - t^2$, the value of the squared interval.

In this sense, then, each observer has his own personal (local) decomposition of space-time into "space" and "time". In this sense, the "obvious" subjective difference between space and time, and their unification into space-time, can co-exist.

Quite generally, given a tensor at p, then any observer who passes through p can, by projecting the indices of that tensor parallel and perpendicular to his four-velocity, express that tensor in terms of tensor all of whose indices are perpendicular to his four-velocity.

We consider one further example of these ideas. Let an observer pass through event p, and also let a particle's world-line pass through p. Let the four-velocities be ξ^a and η^a, respectively. What is the apparent v "velocity of the particle" as seen by the observer? The particle leaves event p to go to event q displaced by η^a from p. Thus, the particle appears to go a spatial distance r (given by (9)) in time t (given by (8)). Hence, the apparent speed of the particle, as seen by our observer, is ($\eta^a \eta_a = -1$)

$$v = \frac{r}{t} = \frac{[(\xi^a \eta_a)^2 - 1]^{1/2}}{(-\xi^a \eta_a)} \tag{10}$$

Note that this speed is always less than one (i.e., less than the speed of light, since our units of speed are light-seconds/second).

With each material particle there is associated a number m called the (rest) *mass* of the particle. The *four-momentum* of the particle is m times the four-velocity of the particle. An observer with four-velocity ξ^a computes $\xi^a p_a$, where p_a is the four-momentum, and calls this the apparent (i.e., to him) energy of the particle. Let's work out the apparent energy in terms of the apparent speed. We have, from (11)

$$\text{Energy} = -\xi^a p_a = -m\, \xi^a \eta_a = \frac{m}{\sqrt{1 - v^2}} \tag{11}$$

This formula is, of course, famous.

8. Acceleration

Some world-lines are more desirable (for an observer to follow) than others. Suppose you decide to jump off the fourth floor of Eckart. Your world-line would look like that in the figure (note the event "jumping", and the event "landing on the sidewalk"). Why is it that one suffers injuries at the event of meeting the sidewalk, rather than, say, at the event of jumping? One would normally say that he underwent a

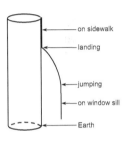

very large acceleration on meeting the sidewalk. In terms of the observer's world-line, this feature is reflected in the fact that his world-line has a "kink" at the event "meeting the sidewalk". Our purpose is to formulate the remarks above quantitatively.

We first remark that, at each event on his world-line, an observer can determine a quantity called his acceleration. Such a determination could, for example, be carried out as follows. Our observer constructs a cubic box inside of which there is a small mass, suspended by springs attached to the faces of the cube. (Such an instrument is called

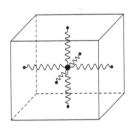

an accelerometer.) At each event on his world-line, our observer can measure the displacement of the small mass from its (normal) position at the center of the cube. This (spatial) displacement defines a vector (in space) for our observer. Hence, from the discussion of Sect. 7, we have a vector perpendicular to the four-velocity of the observer. (By convention, this vector, the acceleration, is defined as minus the displacement of the small mass from the center. Thus, if the cube is accelerated forward, the mass from the center. Thus, if the cube is accelerated forward, the mass would reside near the rear of the cube.)

To summarize, *at each event of his world-line an observer can, by means of a physical measurement, determine a vector A^a orthogonal to* *his four-velocity. This vector is called the acceleration of the observer (or, of the world-line).* (In other words, an observer's acceleration is an absolute, local quantity, measurable without reference to anything external.)

Let us now return to the observer who jumps off Eckart. While standing on the window sill before the jump, his acceleration is 32 feet/sec^2 (in magnitude), directed upward (the mass shifts to the lower half of the cube). After he jumps (i.e., in free fall), his acceleration is zero. This, of course, is just the reverse

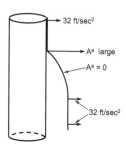

of the usual description (in which standing on the window sill is "not accelerating", and falling is "accelerating at 32 ft/sec^2 is response to the Earth's gravitational field"). The difference is that acceleration (as we have defined it) is not relative to the Earth's surface, or relative to anything. It is the reading from one's cube. One might say that we have "neglected to subtract off the effect of the Earth's gravitational field on the mass inside the cube". That is precisely what we have done... on purpose.

Let us suppose that there were an electric field present, and that our observer were charged. Then he would move in response to this electric field. The mass in the cube would record these accelerations provided that mass were uncharged. If, on the other hand, the mass were also charged (with the same charge/mass ratio as that of the observer holding the cube), then observer and mass in the cube would move together, and the cube would always measure zero acceleration. Why don't we use a charged mass in our cube? Because it is possible to "turn off" electromagnetic forces by making the mass neutral. Gravitational forces, on the other hand, cannot be "turned off". This is the all-inclusiveness of gravitational phenomena – the equivalence principle. Thus, the prescription is that we make the mass inside the cube insensitive to everything we can (i.e., insensitive to everything except gravitation).

We now adopt the attitude that, *whenever a particle's world-line experiences an acceleration, that acceleration must be attributable to some non-gravitational influence.* Thus, the observer standing on the window sill attributes his acceleration (32 ft/sec^2, upward) to the force the window sill is exerting on his feet. After jumping, the window sill no longer exerts that force, and, indeed, the observer measures (with his cube) zero acceleration.

Finally, consider the elevator. Observers inside the elevator cannot tell whether the elevator is sitting on the surface of the Earth, or the elevator is being accelerated through space (at 32 ft/sec^2) by a robe attached to it. But, from the present point of view, their description of the two situations are the same. In both cases, they would say that they are accelerated because of the force the floor of the elevator exerts on their feet. Thus, the equivalence principle (physical indistinguishability of a uniform gravitational field from acceleration) leads, in general relativity, to calling both phenomena the same thing: acceleration.

The discussion above is intended to bring out a very important point of view in general relativity. *An observer does not* (because he cannot, in principle) *attempt to decompose his acceleration into a "real acceleration" and an "apparent acceleration due to gravitational forces on the small mass in his cube".* *These two quantities* (which are distinct in Newtonian gravitation) *are combined into a single, unambiguous quantity – acceleration – measured by the displacement of the small mass inside the cube.*

Note: our units of length are light-seconds (distance light travels in a second). So, 1 ft = 10^{-9} light-seconds. So, acceleration has units light-seconds/second^2 = sec^{-1}. Thus, 32 ft/sec^2 = 3×10^{-8} sec^{-1}.

9. Acceleration and World-line Curvature

We have seen in Sect. 8 that, at each point of a world-line there is associated a vector A^a, the acceleration vector of the world-line, which is orthogonal to the four-velocity at that point. We now want to relate this (physically measured) quantity to some geometrical (mathematically expressed) property of the world-line.

Let us return again to the figure on p. 31. The observer in that figure suffers a very large acceleration at the event "landing on the sidewalk". Is there any way in which this physical fact appears in the "shape" of his world-line at that event? Of course there is: his world-line has a kink at that event. Thus, we expect that acceleration should somehow be related to the extent to which a world-line "bends".

ξ^a at q parallel transported to p

There is a convenient measure of "bending". Let $\gamma: I \to M$ be a timelike curve, parameterized by length. Denote by ξ^a the corresponding tangent vector to this curve. The quantity

$$C^a = \xi^m \nabla_m \xi^a$$

is called the *curvature* of the world-line. Note that the curvature is essentially the derivative of the four-velocity in the direction of the four-velocity. This curvature has the expected geometrical interpretation: it is large for a world-line which "curves sharply", etc. Note also that the curvature of a curve vanishes if and only if that curve is geodesic. Note also that the curvature of a world-line is necessarily orthogonal to the tangent vector to that world-line:

$$\xi_a C^a = \xi_a(\xi^m \nabla_m \xi^a) = \frac{1}{2}\xi^m \nabla_m(\xi_a \xi^a) = \frac{1}{2}\xi^m \nabla_m(-1) = 0.$$

We now introduce the following important assumption: *the* (phys-ically measured) *acceleration of a world-line is equal to the* (geomet-rical) *curvature of that world-line*, $C^a = A^a$. Note firstly that this identification is at least reasonable because both vectors are orthogo-nal to the four-velocity. Furthermore, it is suggested by the first figure of Sect. 8. Finally, the curvature of a world-line is the right sort of quantity to equate to the acceleration, for the former is the derivative of the four-velocity, and so is an "acceleration-like" quantity.

Although these remarks make the assumption plausible, they by no means establish it conclusively. The assumption related space and time measurements (the measurements which give squared intervals, hence the metric, which is needed to find the curvature) to acceleration measurements (by an observer's cube). I know of no "conclusive" argument for it. In fact, one can write down theories in which it does not hold. It is simply an assumption (a very strong assumption with little experimental foundation) of general relativity. Perhaps the brunt of the assumption is its assertion that there is not some additional tensor field on M which combines with curvature of the world-line to produce the acceleration.

We shall hereafter normally use the word "acceleration" in place of "curvature of timelike world-line".

Let a particle of mass m travel on a certain timelike world-line. The quantity mA^a is called the *force* on the particle, F^a. (Newton's law: $F = ma$.) Noting that the four-momentum of the particle, p^a, is given by $m\xi^a$, where ξ^a is the four-velocity of the particle, we have $F^a = a\xi^m \nabla_m \xi^a = m^{-1}p^m \nabla_m p^a$. It is this force which must be ascribed to non-gravitational influences on the particle. We shall obtain examples of such forces (electromagnetic, pressure) shortly.

Note, in particular, that when there is no force on a particle (i.e., when we have a free particle), the world-line of that particle is a geodesic. *A free particle travels on a timelike geodesic.*

Recall the force law in Newtonian gravitation: $\vec{F} = -m\vec{\nabla}\varphi$. Our assumption ($A^a = C^a$) is essentially the relativistic version of this. Think of the metric g_{ab} as analogous to the "gravitational potential", φ. Than C^a involves (in terms of components) the connection, and hence first (coordinate) derivatives of the components of the metric. Thus, in each case, the force is expressed in terms of first derivatives of the potential. (The distinctive feature of general relativity is that g_{ab} has two interpretation: as a geometrical quantity and as a gravitational potential.)

We remark on an obvious consequence of the discussion above. It is common in physics to describe effects on particles in terms of a force field (e.g., electric field, Newtonian gravitational field, etc.). That is, one has a vector field F^a such that, at each point, the force on a particle at that point is F^a. No such force fields are allowed in general relativity. The reason is that the force F^a must be orthogonal to the four-velocity of the particle on which this force acts. If we were given F^a at the beginning, then we could always find some timelike world-line to which p were not orthogonal.

As a final example of these geometrical concepts, we return to the example of several observers passing between a pair of events p and q. In the figure, B is the longest curve. (Recall that a null curve has zero length. In A and C, the four-velocity tends to be "nearer the light cone", and hence A and B have the shorter lengths.) But B is also the curve in which the acceleration is smallest. C appears to go away from B at a large speed, accelerate quickly to turn around, and come back at 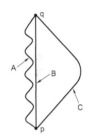 a large speed. On the other hand, A never gets too far from B, but he is on a "bumpy ride", accelerating this way and that. Thus, as a general rule, the observers who endure the most acceleration come back the youngest. A bumpy airplane flight ages you less than a smooth one. If you throw a rock in the air and catch it, the rock has aged less during this exercise than you have (for you were undergoing acceleration: the rock was not).

10. Rotation

There is more than one way for an observer to follow a given world-line. Specifically, an observer can rotate.

Rotation, like acceleration, is an unambiguous, absolute, local, measurable quantity. ("Rotation relative to what?", like "Acceleration relative to what?" is a fruitless question.) This remark is made explicit by the following measuring apparatus.

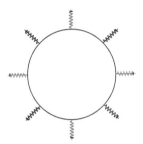

Consider a solid sphere with many spikes sticking out of it. On each spike there is a small bead which is free to slide on the spike. The beads are connected to the sphere by springs so that, at equilibrium, the beads are all at some specified distance (e.g., three inches) from the sphere. Now, when such a sphere is rotated, the beads along the equator of the axis of rotation will move farther out from the sphere (centrifugal force). Thus, one can determine, at any instant, the state of rotation of the sphere by examining the configuration of the beads.

Choose three, orthogonally-directed, spikes on the sphere, and label them u, v, and w. Then, if this sphere is carried along a world-line, the way in which it is carried (regarding rotation) is described by specifying, at each point of the world-line, three vectors, u^a, v^a, w^a, which are orthogonal to each other and to the four-velocity.

In section 9, we related a physical quantity, acceleration, to the curvature of the world-line. We now wish to do a similar thing with rotation. The vectors u^a, v^a, and w^a determine the orientation of the sphere along its world-line. We wish to obtain an expression relating the rate of change of these vectors along the world-line to the rotation measured by the sphere. It suffices to ask the question: what equation must be satisfied by u^a, v^a, and w^a in order that the sphere record zero rotation? Then, the extent to which this equation is not satisfied (for the actual u, v, and w) is a measure of the rotation recorded by the sphere.

We now assert that *the sphere records zero rotation precisely when* u^a, v^a, *and* w^a *satisfy*

$$\xi^m \nabla_m u^a = \xi^a (u_m A^m)$$
$$\xi^m \nabla_m v^a = \xi^a (v_m A^m) \qquad (12)$$
$$\xi^m \nabla_m w^a = \xi^a (w_m A^m)$$

where A^a is the acceleration (curvature) of the world-line. (Eqns. (12) define what is called *Fermi transport*.) We check that this assertion is reasonable. Note, firstly, that Eqns. (12) involve the first derivative of u, v, and w along the world-line. That is to say, they involve the "time rate of change of the orientation of the sphere", which, indeed, is a "rotation-like" quantity. Thus, in particular, if one specifies u^a, v^a, and w^a initially (i.e., at some point on the world-line), then Eqns. (12) determine these vectors uniquely along the entire world-line. (That's correct: if you give me the orientation of the sphere at one time, and I then require that the sphere always record zero rotation, that determines its rotation thereafter.) Note, furthermore, that, if u, v, and w, are initially (i.e., at some point on the world-line) orthogonal, then according to Eqns. (12), they remain orthogonal:

$$\xi^m \nabla_m (u_a v^a) = u_a \xi^m \nabla_m v^a + v^a \xi^m \nabla_m u_a$$
$$= u_a (\xi^a v_m A^m) + v^a (\xi_a u^m A_m) = 0.$$

This, of course, is expected. Finally, note that a vector, transported according to Eqns. (12), remains orthogonal to the four-velocity:

$$\xi^m \nabla_m (u_a \xi^a) = u_a \xi^m \nabla_m \xi^a + \xi^a (\xi^m \nabla_m u_a)$$
$$= u_a (A^a) + \xi^a (\xi_a u_m A^m) = u_a A^a - u_m A^m = 0.$$

Thus, Fermi transport has all the properties one would desire to describe the orientation of a sphere which records zero rotation.

The following remarks are intended to give an intuitive picture of Fermi transport. Consider a circle in the plane. Let p be a point on the circle, and let u be a vector at p which points outward from the circle. Suppose we parallel transport this u around the circle. the resulting vector u is shown in the figure. Note

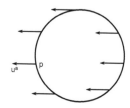

that, while this u is initially orthogonal to the
tangent vector to the circle, this property is not preserved under parallel transport. If, on the other hand, this property is not preserved
under parallel transport. If, on the other hand, v were a vector at p
which sticks out of the paper, the parallel transport of v around the
circle would keep v orthogonal to the tangent vector to the circle. Why
this difference between between u and v? Note that the curvature of
the circle, C, is a vector pointing inward toward the center. Thus,
since the circle "curves in (minus) the direction of u", and u does not
remain orthogonal to the tangent to the circle. Since the curvature of
the circle is orthogonal to v, v does remain orthogonal to the tangent
vector.

Initially, in Fermi transport one "parallel transports the vector a
little way along the curve, then projects orthogonal to the tangent
vector, parallel transports a little further, then projects again, etc."
Under "parallel transport for a little way" the vector "loses its orthogonality with the tangent vector" by an amount determined by uA (as
we saw above; u A is nonzero. and u loses orthogonality under parallel
transport). Thus, to "restore orthogonality" one must correct parallel
transport by an amount involving uA. Thus, the term on the right in
(12). The appears on the right because we want to "correct parallel
transport of u by a multiple of the tangent vector". In other words, we
want to project orthogonally to (an operation which is accomplished
by adding an appropriate multiple of the tangent vector). In short,
Fermi transport is parallel transport with a correction term to keep
the transported vector orthogonal to the four-velocity.

Note, in particular, that for a geodesic (i.e., for A = 0) Fermi
transport and parallel transport coincide.

Finally, the figure on the right represents
the result of Fermi transport of u (given at p)
around the circle. that this is the right answer
should be clear from the remark above. For
v (the vector which sticks out of the paper),
Fermi transport and parallel transport coincide
(for vA = 0. See (12).) (In general, parallel
transport about a closed curve leaves a vector
unchanged only when the space is flat. Fermi
transport about a closed curve does not leave a vector unchanged -
even in flat space.)

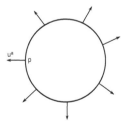

.

11. Dust

The source of a gravitational field is matter. In Newtonian theory, this fact is expressed be the equation $\nabla^2\varphi = 4G\rho$. Thus, all the various types of matter have in common, in Newtonian theory, the feature that they determine a ρ, the matter density, which generates gravitation. In general relativity, matter generates gravitational field via a certain tensor field T_{ab}, called the *stress-energy tensor* of the matter. (So, T_{ab} is analogous to the Newtonian ρ.) We now begin our study of various types of matter to see how each, ultimately, defines a stress-energy tensor.

Consider a cloud of dust, i.e., an assembly of a large number of very small, very light particles which do not directly interact with each other. In terms of space-time, we would represent this dust cloud by drawing the world-lines of all the dust particles. This congruence of time-like curves would fill a certain region of space-time.

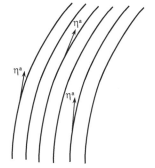

The idea is to describe this situation in terms of certain tensor fields in space-time. At each event of space-time inside the dust cloud, let η^a be the four-velocity of the dust particle there. In this way, we obtain a unit $(\eta^a\eta_a = -1)$ vector field η^a in the region of space-time occupied by the dust. This η^a is called the (four-) *velocity field* of the dust. This is the first of two tensor fields describing the dust.

The second is the *density*, ρ, of the dust. At a point p of the dust cloud, draw a small 3-surface S orthogonal to η^a. Denote by V the volume (in $\sec^3$) of this surface. Then ρ is defined as Nm/V, where N is the number of dust particle world-lines which pass through S, and m is the mass of each dust particle. (So, ρ has units grams/$\sec^3$.) Note that this ρ is just the usual "mass density" as would be determined by an observer traveling along a world-line of the dust. This ρ is a scalar field in the region occupied by the dust. (More precisely, we are thinking of a limit in which the number of dust particles increases, and the mass of each one approaches zero, such that ρ remains finite in the limit.)

Note: density of water $= 3 \times 10^{31}$ g/sec^3.

We assume that ρ and η^a are smooth fields. (Rule of thumb: Whenever smoothness is meaningful for physical quantity, assume it.)

What equations would we expect the fields ρ and η^a to satisfy? The first expresses the fact that the dust particles are non interacting, i.e., that they are free particles. The corresponding equation, clearly, is $\eta^m \nabla_m \eta^a = 0$. That is, the dust particles travel along geodesics. The second equation expresses the conservation of energy (or, since all particles have a fixed mass m, of dust particles). Consider a tubular volume V whose sides are parallel to η^a and whose faces are orthogonal to η^a. The total amount of mass (in dust) entering this tune is

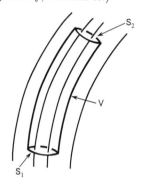

$$\int_{S_1} (\rho \eta_a) dS^a$$

where S_1 is the bottom face of the tube, and dS^a is the volume element on this face (directed outward). Similarly, the total mass in dust leaving the tube is

$$-\int_{S_2} (\rho \eta_a) dS^a$$

Conversation of dust particles requires

$$\left(\int_{S_1} + \int_{S_2} \right) (\rho \eta_a) dS^a = 0$$

or, what is the same thing,

$$\int_{2\nabla} (\rho \eta_a) dS^a = 0$$

In the second surface integral, we have included the sides of the tube, but these make no contribution, since $\eta_a dS^a = 0$ there. By Gauss' law, we now have

$$\int_{\nabla} \nabla_a(\rho\eta_a)\mathrm{d}V = 0$$

Finally, since the volume V is arbitrary, we have $\nabla_a(\rho\eta^a) = 0$. Intuitively, this equation states that, when the world-line of the dust-particles draw closer together, the density ρ increases by a corresponding amount. (Write the equation in the form $\eta^a\nabla_a\rho = -\rho\nabla_a\eta^a$ to see this.) This, of course, in what we would expect to happen from conservation of dust particles.

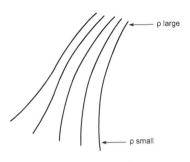

All the discussion above can be regarded as merely motivation. When we say "dust", we mean merely a pair of fields, and , satisfying

$$\eta^m\nabla_m\eta^a = 0 \qquad (13)$$

$$\nabla_a(\rho\eta^a) = 0 \qquad (14)$$

(and, of course, $\eta^a\eta_a = -1$). One would normally also require $\rho \geq 0$, i.e., that the dust particles have positive mass.) Eqns. (13) and (14) can be written as one. Multiplying (13) by ρ, and (14) by η^a, and adding:

$$\rho\eta^m\nabla_m\eta^a + \eta^a\nabla_m(\rho\eta^m) = 0 \qquad (15)$$

Conversely, contracting (15) with η_a (using $\eta_a A^a = 0$), we obtain (14). Then (15) gives (13). Thus, Eqn.(15) is completely equivalent to (13) and (14). Finally, note that (15) can be written

$$\nabla_b T^{ab} = 0 \qquad (16)$$

where we have set $T^{ab} = \rho\eta^a\eta^b$. This T^{ab} is called the *stress-energy* of the dust. The behavior of the dust is then completely described by (16), which is called the *conservative equation*.

We next give direct physical interpretation of T^{ab}. Let an observer pass through our dust cloud, and let his four-velocity be η^a. We claim that the quantity $-\eta_b T^{ab}$ is the four-momentum density of the dust, as seen by this observer. (So, in particular, $\xi_a\xi_b T^{ab}$ is the observed energy density.) Suppose first that $\xi^a = \eta^a$. Then $-\eta_b T^{ab} = \rho\eta^a$. Evidently (from the definition of ρ), this is the four-momentum density of the dust as seen by an observer moving with the dust. (Recall that $m\eta^a$ is the four-momentum of a single dust particle. So, $\rho\eta^a = (N/V)m\eta^a$ is the four-momentum density as seen by this observer.) Next, suppose general ξ^a Then $-\eta_b T^{ab} = (-\xi_b\eta^b)\rho\eta^a$. The second factor on the right is (as noted above) the four-momentum density seen by an observer following the dust. We shall show below that $(-\xi_b\eta^b)^{-1}$ is precisely the ratio of the density of dust particles seen by the ξ-observer to the density seen by the η-observer. This having been done, the interpretation claimed above for $-\xi_b T^{ab}$ will be clear.

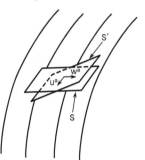

Fix an event p on the world-line of our observer. Let S be a small cubical space-like 3-surface at p which is orthogonal to η^a. By "tipping" the boundary of S, construct a second cubical spacelike 3-surface S' at p, this one orthogonal to ξ^a. Clearly, the same number of dust trajectories pass through S and S'. Thus, we have only to show that $V'/V = (-\xi_a\eta^a)^{-1}$, where V' and V are the areas (in sec^3) of S' and S, respectively. Let the sides of the cube S be described by unit vectors u^a, v^a and w^a at p, where these vectors are orthogonal to each other and to η^a. Furthermore, let w^a lie in the (2-dimensional) plane of ξ^a and η^a. Then the sides of S' are described by the vectors u^a, v^a, and $w^a - \eta^a(\xi^m w_m)(\xi^n\eta_n)$ i.e., we add the appropriate multiple of η^a to w^a so the result will be orthogonal to ξ^a). But, since w^a is unit, orthogonal to η^a and a linear combination of ξ^a and η^a, we have $w^a = ((\xi^m\eta_m)^2 - 1)^{-1/2}(\xi^a + (\xi^m\eta_m)\eta^a)$. Hence, $\xi^m w_m = ((\xi^m\eta_m)^2 - 1)^{1/2}$. We conclude that the edges, of S' are described by u^a, v^a, and $w^a - \eta^a(\xi^a\eta_m)^{-1}((\xi^n\eta_n)^2 - 1)^{1/2} = w'^a$. We now have three vectors describing the edges of S'. Clearly, $V'/V = (w'^a w'_a)^{1/2} = (-\xi^n\eta_n)^{-1}$. This completes the proof. (This argument is well-known in special relativity. it is called the "transformation law for a density under a boost", and the formula is explained by saying that "one of the three spatial dimensions suffers a Lorentz contraction".)

It is true quite generally (as shown explicitly above) that $-\xi_b T^{ab}$ *can be interpreted as the four-momentum density of the matter described by T^{ab}, according to an observer with four-velocity ξ_b.* Thus, we can think of T^{ab} as having two indices because "one will be a four-momentum index, while the other must be contracted with ξ_b in order that the correct density transformation between observers obtains".

In astrophysics, the stars in a galaxy, or the galaxies in the universe, are often treated as "dust particles". Dust is one of the simplest sources.

12. Perfect fluids

The second type of matter we shall discuss are the perfect fluids (e.g., water, air). Dust, as we shall see, may be regarded as a special case of perfect fluid.

Consider a fluid residing in our space-time. As usual, we want a description in terms of tensor fields. Imagine tagging in some way a small volume of fluid. In space-time, this volume would have some world-lines. Let η^a be the unit tangent vector to this world-line. Repeating, at each event of space-time in the fluid, we obtain a unit timelike vector field η^a. This is called the (four-) *velocity field* of the fluid. At each event in the fluid, an observer passing though that event with four-velocity η^a could measure the density (g/sec^3) of the fluid. Thus, we have the *density* ρ of the fluid a scalar field in space-time. Finally, an observer with four-velocity η^a could determine a quantity p, the *pressure* of the fluid. Of course, p is a scalar field in space-time. (The units of pressure are g/sec^3. Thus, 1 dyne /cm^2 = 3 × 10^{10} g/sec^3.)

Thus, a perfect fluid is defined by three fields, η^a, ρ, and p, in space-time. One would normally require that ρ be non-negative. (On the other hand, many fluids, e.g., water, can exert negative pressure.) The specification of what fluid is under consideration is carried out by giving the pressure p as a function of ρ, $p(\rho)$. (For example, if you say what you want the density of air to be, I can tell you what pressure must be exerted on a sample of air to achieve that density. The function $p(\rho)$ would, of course, be quite different for water.) This $p(\rho)$ is called the *equation of state* of the fluid. (Of course, one could consider more complicated fluids. One could, for example, introduce temperature, entropy density, etc., and write more complicated equations of state. While these more complicated fluids are important for certain applications, their introduction is straightforward, and they add nothing fundamental to the study of general relativity. We shall not consider then.)

We could now proceed with fluids as we did before with dust. We could agree on some reasonable equations to be satisfied by η^a, ρ , and p. Instead, we shall simply write down the stress-energy tensor for a perfect fluid, and then verify that its conservation leads to reasonable physics.

The stress-energy tensor for a perfect fluid is given by

$$T^{ab} = (\rho + p)\ \eta^a \eta^b + p\ g^{ab}.$$

It follows immediately that dust is a special case of a perfect fluid – the case with equation of state $p(\rho) = 0$. The conservation of stress-energy, $\nabla_b T^{ab} = 0$, gives for a perfect fluid:

$$(\rho + p)\eta^b \nabla_b \eta^a + (g^{ab} + \eta^a \eta^b)\nabla_b p + \eta^a p \nabla_b \eta^b + \eta^a \nabla_b(\rho \eta^b) = 0 \quad (17)$$

In order to extract some physics from (17), we project parallel and orthogonal to η^a. (This procedure is useful, as a general rule, when dealing with fluids.) Contracting (17) with η^a, we have

$$\nabla_b(\rho \eta^b) = -p\nabla_b \eta^b. \tag{18}$$

For dust, the right side of (18) vanishes. In this case, our interpretation of (18) was that it gives conservation of mass (for the dust). Thus, the presence of a non vanishing right side of (18) indicates that mass is no longer conserved. How shall we interpret this physically? Consider (in Newtonian physics) a sample of gas of volume V, and at pressure p. Suppose this gas is allowed to expand slightly, so its volume becomes $V + dV$. Then, in this process, an energy pdV is extracted from the gas. (Let the gas be in a cylinder with a piston of area A. Then $dV = Adl$, where dl is the distance the piston moves. The pressure force on the piston is pA. Hence, the energy extracted is force $\times$ distance $pA \times dl = pdV$.) Recall that $\nabla_b \eta^b$ represents the rate at which a small volume of fluid is expanding. The interpretation of (18) is now clear: it states that the energy pumped into the fluid (by pressure acting against volume changes) goes into increasing the mass density of the fluid. (If you compress a sample of gas, it gets heavier because of the energy you are putting into it. $E = mc^2$, and all that.) Thus, Eqn. (18) is conservation of mass-energy.

The other half of Eqn. (17) is obtained by projecting orthogonal to η_a. This is accomplished by contracting it with one index of the projection operator, $h_{ac} = g_{ac} + \eta_a \eta_c$, orthogonal to η^a. Thus, we obtain

$$(\rho + p)\ \eta^b \nabla_b \eta^a = -(g^{ab} + \eta^a \eta^b)\nabla_b p. \tag{19}$$

Again, in the case of dust, the right side of (19) vanishes, and the equation then represents the statement that the dust particles are free particles. Hence, in the present case, we wish to interpret the right side of (19) as a force on small volume elements of the fluid. Note that the right side of (19) can be written $h^{am}\nabla_m p$, where h^{am} is the projection operator orthogonal to the fluid. In other words, the right side of (19) represents a pressure force on fluid elements. Recall that, in Newtonian mechanics, the gradient of the pressure gives a difference in the force on opposite faces of a small fluid element, and hence a net force on that element. The situation is the same here, except that $\nabla_b p$ is the gradient in space-time, and this gradient must be projected orthogonal to η^a to obtain the force. Note also that the coefficient of the acceleration on the left in (19) is $(\rho + p)$. This quantity thus represents the "effective inertial mass density" of the fluid.

We have in (19) an example of a force in general relativity. One begins with a vector field (in the example above, $\nabla_b p$). Then, the force on a particle is obtained by projecting that vector orthogonal to the four-velocity of the particle.

To summarize, a perfect fluid is characterized by fields, η^a, ρ, and p, which define a stress-energy T^{ab}. The component of $\nabla_b T^{ab} = 0$ parallel to η^a gives conservation of mass-energy, the component orthogonal to η^a conservation of momentum.

13. Electromagnetic Fields: Decomposition by an Observer

We now begin our treatment of the third important type of "matter": electromagnetic fields.

An *electromagnetic field* is a (smooth) antisymmetric tensor field, F_{ab} ($= F_{[ab]}$), on space-time M. An individual observer resolves this single object – the electromagnetic field – into the separate electric and magnetic fields seen by him. Our first task is to see how this resolution comes about. Let our observer have four-velocity ξ^a. Recall that $h_{ab} = g_{ab} + \xi_a \xi_b$ is the projection operator orthogonal to ξ^a. That is, if w^b is w is any vector (at a point of the observer's world-line), then $h^a{}_b w^v$ is the component of w^b orthogonal to ξ^a. (Proof: For any w^b we have $w^a = h^a{}_b w^b - \xi^a(\xi^m w_m)$. But the second term on the right is the component of w^b parallel to ξ^a.) Thus, in particular we have $h^a{}_m h^m{}_b = h^a{}_b$ (the component orthogonal to ξ of the component orthogonal to ξ of a vector is the component orthogonal to ξ of that vector), and $\xi^b h^a{}_b = 0$ (the component of ξ^b orthogonal to ξ is zero).

To resolve F_{ab} into "spatial tensors" for our observer, we project the indices of F_{ab} parallel and orthogonal to ξ^a. Thus, we obtain four tensors: $F_{mn}\xi^m\xi^n$, $F_{mn}\xi^m h^n{}_a$, $F_{mn}h^m{}_a\xi^n$, and $F_{mn}h^m{}_a h^n{}_b$. Since F_{ab} in antisymmetric, the first vanishes, and the second equals minus the third. The third is called the *electric field*:

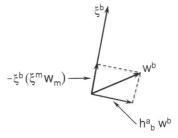

$$E^a = F_{ab}\,\xi^b. \qquad (20)$$

(Note that $F_{am}\,\xi^m = F_{mn}\xi^m h^n{}_a$) Note that the determination of

41

an electric field from the electromagnetic field involves a choice of an observer (i.e., of his four-velocity). The remaining piece of F_{ab} is $F_{mn}\, h^m{}_a\, h^n{}_b$, a spatial, antisymmetric tensor. To express this piece more simply, let ϵ_{abcd} be the alternating tensor on M (i.e., the (unique up to sign) tensor field defined by $\epsilon_{abcd} = \epsilon_{[abcd]}$ and $\epsilon_{abcd}\epsilon^{abcd} = -24$). Then $\epsilon_{abc} = \epsilon_{abcd}\, \xi^d$ is the spatial alternating tensor as seen by our observer (i,e., it satisfies $\epsilon_{abc} = \epsilon_{[abc]}$, $\epsilon_{abc}\epsilon^{abc} = 6$, and $\xi^a\, \epsilon_{abc} = 0$).

A spatial anti symmetric W_{ab} for our observer (i.e,. satisfying $W_{ab} = W_{[ab]}$, and $\xi^a\, W_{ab} = 0$) is then completely described by the corresponding vector, $\epsilon^{abc}\, W_{bc} = W^a$. Note that W^a is also a spatial vector (for our observer), and that W_{ab} can be recovered from W^c: $W_{ab} = \frac{1}{2}\epsilon_{abc}\, W^c$. The spatial vector which gives $F_{mn}\, h^m{}_a\, h^n{}_b$ is thus

$$B_a = \frac{1}{2}\, \epsilon_{abcd}\, \xi^b\, F^{cd}. \tag{21}$$

This B_a is called the *magnetic field*.

To summarize, the whole story is described by a single tensor field, the electromagnetic field. An observer resolves this F_{ab} using his four-velocity, into a pair of vectors, E_a and B_a, orthogonal to his four-velocity. The E_a and B_a depend on the observer. (In terms of components, the six independent components of F_{ab} become three in E_a and three in B_a.)

We now reverse the procedure of Eqns. (20) and (21): we express F_{ab} in terms of E_a and B_a. The equation is

$$F_{ab} = 2\xi_{[a}E_{b]} - \epsilon_{abcd}\, \xi^c\, B^d. \tag{22}$$

To verify this one substitutes (22) into (20), and (22) into (21), to obtain identities.

An electromagnetic field produces a force on a charged particle. Consider a particle with charge e, and let η^a be the four-velocity of the particle. Then the force on this particle is the right side of:

$$m\, \eta^b\, \nabla_b\, \eta^a = e\, F^{ab}\, \eta_b. \tag{23}$$

Thus, Eqn. (23) is the equation of the motion of the particle (m is its mass). Note that the right side is orthogonal to η_a (as we would demand, since the left side automatically is), since F_{ab} is skew. Eqn. (23) gives immediately the equation of motion for charged dust. Suppose each particle of the dust has charge e, and let μ be the charge density of the dust (defined exactly as the mass density. Thus, in particular $\rho/\mu = m/e$.) Then, in this case, Eqn. (13) would be replaced by

$$\rho \, \eta^b \, \nabla_b \, \eta^a = e \, F^{ab} \, \eta_b. \tag{24}$$

Our next objective is to recover from (23), the usual classical formula for the force on a charged particle. Through a point of space-time, let there pass an observer with four-velocity ξ^a, and a charged particle with four-velocity η^a. We suppose (to simplify the discussion) that η^a differs infinitesimally from ξ^a, i.e., we set $\eta^a = \xi^a + v^a$, where v^a is small. (It follows from the fact that η^a and ξ^a are unit vectors that $\xi^a \, v_a = 0$.) This v^a is the velocity of the charged particle, as seen by the observer. Now, our observer decomposes F_{ab} into E_a and B_a as described above. Hence, he can express the force on the particle in terms of E_a, B_a, and v_a. Unfortunately, this force, $e \, F_{ab} \, \eta^b$, is orthogonal to η^a (the four-velocity of the particle), and not to ξ^a (the observer's four-velocity). However, we can project it orthogonal to the observer's four-velocity using $h_{ab} = g_{ab} + \xi_a \xi_b$. Thus, the force is given by (first order in v^a)

$$e \, h_a^m \, \eta^n F_{mn} = e(\delta_a^m + \xi_a \, \xi^m)(\xi^n + v^n)$$

$$(e \, E_{[m} \, \xi_{n]} - \epsilon_{mncd} \, \xi^c \, B^d)$$

$$= e \, E_a + e \, \epsilon_{abc} \, v^b \, B^c \quad (\text{where} \quad \epsilon_{abc} = \epsilon_{abcd} \, \xi^d)$$

In standard three-dimensional vector notation, this equation would be written $\vec{F} = e \, \vec{E} + e \, \vec{v} \times \vec{B}$, the usual force equation.

In everyday life in general (or special) relativity, one never uses the decompositions of things as seen by an observer. These are given here merely for motivation – as an opportunity to recover some familiar formulae. Thus, eliminating all this motivation, there are only two statements in this entire section. The electromagnetic field is a skew tensor field, F_{ab}, on space-time. The force on a particle of charge e is given by the right side of Eqn. (23).

14. Maxwell's equation

We have now introduced electromagnetic fields, and the equation (23) for the action of this field on charged particles. the next step is to describe the effects of charges on the fields. These, Maxwell's equations, express the production of the electromagnetic fields by charges and currents.

Maxwell's equations are

$$\nabla^{[a} F^{bc]} = 0. \tag{25}$$

$$\nabla_b F^{ab} = J^a. \tag{26}$$

Here, F_{ab} is the electromagnetic field, and J_a is a certain vector field in space-time called the *charge-current density*. This charge current density is obtained from the field describing the charged matter. We begin with the physical interpretation of J_a. Let an observer have four-velocity ξ^a. Then $(-J_a \, \xi^a)$ is the *charge-density* as seen by our observer, while $h_a{}^m J_m$ is the *current-density* as seen by this observer. For example, for charged dust, the charge-current density is given by

$$J^a = \mu \, \eta^a,$$

where μ is the charge-density of the dust. For example, for an observer with four-velocity η^a – i.e., an observer traveling with the dust – the charge-density he sees is $(-J_a \, \eta^a) = -\mu \, \eta_a \eta^a = \mu$. He sees zero current-density (as one would expect, he moves with the dust. On the other hand, an observer with four-velocity different from that of the dust sees a current-density (physically, he sees all those charged particles rushing by).

An important equation on J_a follows immediately from (26). Taking the divergence of this equation, we have

$$\nabla_a \, J^a = \nabla_a \nabla_b \, F^{ab}.$$

But, from the definition of the Riemann tensor (commuting derivatives),

$$\nabla_a \nabla_b \ F^{cd} = -\nabla_b \nabla_a \ F^{cd} = R_{ab}{}^c{}_m \ F^{md} + R_{ab}{}^d{}_m \ F^{cm}. \qquad (27)$$

Contracting (27) over "a" and "c", and over "b" and "d", using the fact that F_{ab} is antisymmetric, we see that the right side of the preceding equation vanishes. That is, $\nabla_a J^a = 0$. What this means should be clear from Sect. 11. It is conservation of charge. Thus, the source for the electromagnetic field, the charge-currant density, is divergence-free, expressing conservation of charge.

Finally, we verify that (25) and (26), when expressed in terms of E_a and B_a (as determined by some observer), become the usual formulae for Maxwell's equations. Unfortunately, Eqns. (25) and (26) involve derivatives; hence, they are difficult to express at a point in terms of the observations of a single observer. We therefore introduce a field of observers.

Let our space-time be that of special relativity (i.e., let the Riemann tensor vanish), and let ξ^a be a constant ($\nabla_a \ \xi^a = 0$, unit, timelike vector field in space-time. (There exists no such vector field in general in general relativity. Proof: $0 = \nabla_{[a}(\nabla_{b]} \ \xi^c) = \nabla_{[a}\nabla_{b]} \ \xi^c = \frac{1}{2}R_{ab}{}^c{}_d \ \xi^d$.) We can now carry out the decomposition of F_{ab} into E_a and B_a at each point of space-time. We also decompose our derivative operator: let a dot denote $\xi^a \ \nabla_a$ (the time derivative), and let $D_a = h_a{}^b\nabla_b$ (the spatial derivative). Thus, we now have a special symbol for the spatial and temporal parts (with respect to our field of observers, i.e., with respect to ξ^a) of every quantity appearing in (25) and (26).

Contacting Eqns. (26) with ξ^a, we have $\xi_a\nabla_b F^{ab} = \xi_a J^a$. But $\xi_a\nabla_b F^{ab} = \nabla_b(\xi_a F^{ab}) = -\nabla_b E^b = -h^{ab}\nabla_a E_b = -D_b E^b$. Furthermore, $\xi_a J^a = -\kappa$, where κ is the charge-density (as seen by our field of observers). Hence, Eqn. (26), contracted with ξ_a, yields

$$D_a E^a = \kappa. \qquad (28)$$

This is the first (of four) Maxwell equations. The second Maxwell equation is obtained by projecting the index of Eqn. (26) orthogonally to ξ^a: $h_{ac} \ \nabla_b F^{cb} = h_{ac}J^c$. But $h_{ac}J^c = j_a$ the current-density (as seen by our field of observers). Furthermore,

$$h_{ac}\nabla_b F^{cb} = \nabla_b(h_{ac}F^{cb}) = \nabla_b(-E_a\xi^b - \epsilon_a{}^{bcd}\xi_c B_d)$$
$$= -\xi^b\nabla_b E_a - \epsilon_{abcd}\xi^c\nabla^b B^d = -\xi^b\nabla_b E_a + \epsilon_{abc}D^b B^c.$$

Hence, projecting the index of (26) orthogonally to ξ^a, we obtain

$$\dot{E}_a - \epsilon_{abc}D^b B^c = -j_a. \tag{29}$$

In vector notation, the second term on the left would be written $-\vec{\nabla} \times \vec{B}$. Eqn. (29) is the second Maxwell equation.

The final two Maxwell equations come from (25). The calculation is almost identical to that above, so we do not give it in detail. Write Eqn. (25) in the form $\epsilon_{abcd}\nabla^b F^{cd} = 0$. Contracting with ξ_a, and projecting the index "a" orthogonally to ξ_a, we obtain, respectively,

$$D_a B^a = 0. \tag{30}$$

$$\dot{B}_a + \epsilon_{abc}D^b E^c = 0. \tag{31}$$

Eqns. (28), (29), (30), and (31) will be recognized as Maxwell's equations written out in "space and time" form. We have shown that this usual form for the equations follows from the much simpler ("space-time") equations (25) and (26).

Once again, it is true that almost everything in this section (in particular, everything involving space and time decompositions relative to an observer) can be regarded as merely motivation for what is a very simple situation in terms of space-time. Eliminating all motivation, what remains is the following. There is a vector field J_a in space-time, the charge current-density. This vector field acts as a source for electromagnetic field, in the sense that F_{ab} is required to satisfy Maxwell's equation, (25) and (26). It follows, in particular, from (26) that the divergence of J^a vanishes. (We shall never again use Eqns. (28), (29), (30), and (31).)

15. Stress-Energy of the Electromagnetic Field

We have now written the complete set of equations describing the electromagnetic field. These give the effect of an electromagnetic field on charged matter (Eqn. (23)), and the effect of charged matter on the electromagnetic field (Eqns. (25) and (26)). We have seen in the cases of dust and fluids that the fields which describe the matter can be combined to give a stress-energy tensor, T^{ab} , which is conserved, and which is such that $\xi_a \xi_b T^{ab}$ is the energy density of the matter as seen by an observer with four-velocity ξ^a. We now show that an electromagnetic field defines a stress-energy with similar properties.

The stress-energy tensor of the electromagnetic field is given by the following equation:

$$T^{ab} = F^{am} F^b_{\ m} - \frac{1}{4} g^{ab} F^{mn} F_{mn}. \tag{32}$$

As usual, we begin by seeing what (32) means for our favorite observer – the one with four-velocity ξ^a. First note that, substituting (22), we have

$$
\begin{aligned}
F^{mn} F_{mn} &= (2\xi^{[m} E^{n]} - \epsilon^{mncd} \xi_c B_d)(2\xi_m E_n - \epsilon_{mnrs} \xi^r B^s) \\
&= 2(\xi^m \xi_m)(E^n E_n) - 2(\xi^c \xi_c) B^d B_d \\
&= 2(B^c B_c - E^c E_c),
\end{aligned}
$$

where we have used $\epsilon_{abmn} \epsilon^{cdmn} = -4\delta_a^{\ [c} \delta_b^{\ d]}$. Hence, from (32),

49

$$\xi_a \xi_b T^{ab} = (\xi_a F^{am})(\xi_b F^b_{\ m} - \frac{1}{4}\xi^a \xi_a F^{mn} F_{mn}$$

$$= E^m E_m + \frac{1}{2}(B^m B_m - E^m E_m) \qquad (33)$$

$$= \frac{1}{2}(E^m E_m + B^m B_m)$$

Similarly,

$$-h_{ac}\xi_d T^{cd} = -h_{ac}\xi_d[F^{cm} F^d_{\ m} - \frac{1}{4}g^{cd} F^{mn} F_{mn}]$$

$$= -h_{ac}\xi_d F^{cm} F^d_{\ m} = h_{ac} E_m F^{cm}$$

$$= h_{ac} E_m[2\xi^{[c} E^{m]} - \epsilon^{cmrs}\xi_r B_s] \qquad (34)$$

$$= -h_{ac} E_m \epsilon^{cmrs}\xi_r B_s = \epsilon_{abcd} E^b B^c \xi^d$$

$$= \epsilon_{abc} E^b B^c.$$

The right sides of (33) and (34) will be recognized from classical electromagnetism. The right side of (33) is what is called the energy density of the electromagnetic field. The right side of (34), written $\vec{E} \times \vec{B}$ in vector notation, is called the Poynting vector, and is interpreted as the momentum density of the electromagnetic field.

Thus, if we take for (32) the stress-energy of the electromagnetic field, we obtain the usual conclusions: $-T^{ab}\xi_b$ is the four-momentum density of the electromagnetic field, as seen by an observer with four-velocity ξ_b.

Note that T^{ab} in (32) is symmetric and trace-free (i.e., $T^{ab} = T^{(ab)}$, and $T^m_{\ m} = 0$).

As a consequence of the equations satisfied by the fields describing dust, the stress-energy of dust is conserved. As a consequence of the equations satisfied by the fields describing a perfect fluid, the stress-energy of the fluid is conserved. Is the stress-energy of the electromagnetic field, (32), conserved? We find out:

$$\nabla_b T^{ab} = F^{am}\nabla_b F^b_{\ m} + F^b_{\ m}\nabla_b F^{am} - \frac{1}{2}F_{mn}\nabla^a F^{mn}$$

$$= F^{am}\nabla_b F^b_{\ m} - \frac{3}{2}F_{mn}\nabla^{[a} F^{mn]} \qquad (35)$$

$$= -F^{am} J_m.$$

Where, in the last step, we have used Maxwell's equations, (25) and (26).

Consider first the case when the charge-current J^a vanishes. Then, from (35), we indeed have $\nabla_b T^{ab} = 0$. But, when J^a is nonzero, we do not in general have conservation of the electromagnetic stress-energy. What is going on? Recall that the conservation of stress-energy represents, physically, the local conservation of energy and momentum. An electromagnetic field without sources has an energy and momentum (according to an observer) which is conserved. But, when we have charged particles around, the electromagnetic field has the possibility of transferring energy and momentum to these particles. Hence, it should not be the stress-energy of the electromagnetic field alone which is conserved. Instead, it should be the total stress-energy – the sum of that due to the fields and that due to the sources – which is conserved.

We illustrate the remarks above for the case of charged dust and electromagnetic fields. The fields are the four-velocity of the dust, η^a, the mass density of the dust, ρ, the charge density of the dust, μ, and the electromagnetic field, F_{ab}. These fields satisfy conservation of mass for the dust ($\nabla_a(\rho\eta^a) = 0$), equation of motion for dust particles (Eqn. (24)), and Maxwell's equations (Eqns. (25) and (26)). For the stress-energy of this system, we take the sum of the dust and electromagnetic contributions:

$$T^{ab} = \rho\eta^a\eta^b + F^{am}F^b{}_m - \frac{1}{4}g^{ab}F^{mn}F_{mn}.$$

Hence,

$$\nabla_b T^{ab} = \eta^a\nabla_b(\rho\eta^b) + \rho\eta^b\nabla_b\eta^a + F^{am}\nabla_b F^b{}_m - \frac{3}{2}F_{mn}\nabla^{[a}F^{mn]}.$$

The first term on the right vanishes by conservation of mass for the dust. The second term cancels the third by (24) and (26). The fourth term vanishes by (25). Hence, we obtain $\nabla_b T^{ab} = 0$, as expected.

Once again, this entire section can be regarded as consisting of motivation, except for one point. The stress-energy of the electromagnetic field is given by Eqn. (32).

16. Stress-Energy: A Summary

There are other things in the world besides dust, fluids, and electro-magnetic fields. For example, there are solids (matter with internal stresses), Dirac fields (which describe electrons), etc. In the preced-ing five sections, we have treated just three cases – dust, fluids and electromagnetism – in detail. (These three, at least in part because of their simplicity, are the most common.) Certain broad features of the discussion of the last five sections are, it is assumed, valid for all matter and fields which occur in Nature. For all matter and all fields which have been so far investigated, these broad features have been found to hold. As a summary, we now list these features.

Each type of matter is described by a certain collection of smooth tensor fields in space-time. These fields satisfy certain differential equations. The equations for the fields of one type of matter may involve the fields of another type of matter. In this case, the two types of matter are considered as interacting with one another. For each type of matter, there is an expression for a tensor field T^{ab} in terms of the fields of that matter. This T^{ab}, the *stress-energy* (of that type of matter), is symmetric. Now suppose we assemble a list of various type of matter, where this list has the following property: the equa-tions for all the fields for all the types of matter in this list involve only the fields for the type of matter in the list. (Physically, suppose that each type of matter in the list interacts only with types of matter which appear in the list.) Then T^{ab}, the sum of the stress-energies of all the types of matter in the list, is the total stress-energy of this system of interacting types of matter. Then $\nabla_b T^{ab} = 0$, i.e., the total stress-energy is conserved.

Let an observer have four-velocity ξ^a. Let T^{ab} be the stress-energy of one particular type of matter. Our observer interprets $-\xi_b T^{ab}$ as the four-momentum density of this type of matter, so, in particular, he interprets $\xi_a \xi_b T^{ab}$ as the mass (energy) density of that type of matter. This observer regards conservation of the total stress-energy (of some interacting types of matter) as representing local conservation

of energy and momentum. That the stress-energy of each individual
type of matter is not conserved is no surprise to him. He says that the
various types of matter exchange energy and momentum (from one
type to another) through their interaction.

17. The Optical Limit

It is known that light and electromagnetic fields are by no means disjoint phenomena in physics. In fact, it is known that what we call light is merely waves in the electromagnetic fields. Thus, in particular, all the properties of light follow already from Maxwell's equations ((25) and (26)). Associated with light are diffraction phenomena – phenomena which become important when one looks at distance comparable with the wavelength of the light. In the limit of very small wavelength (i.e., when the wavelength is much smaller than all other relevant distances), the behavior of light is that described by geometrical optics (i.e., light goes in straight lines, etc.). We shall call this (short wavelength) limit in the *optical limit*.

In Sect. 3, we used light to obtain the light-cone structure of space-time. By "light" in that section, we meant light (i.e., electromagnetic waves) in this optical limit. From the way we defined things in Sect. 3, it follows that light rays travel on null curves. But now we have Maxwell's equation. We now ask: is the behavior of electromagnetic waves (as determined by Maxwell's equations) such that, in the optical limit, these waves (light rays) do indeed travel on null curves? We shall here not only answer this question affirmatively, but, furthermore, obtain additional properties of light rays. We shall here take the optical limit of Maxwell's equations.

We consider an electronic field of the form

$$F_{ab} = \overline{F}_{ab} \, \sin(\alpha\varphi) \tag{36}$$

In (36) $\overline{F}_{ab}$ is an antisymmetric tensor field in space-time which vanishes outside of some tube in space-time. This ensures that, in terms of the "space and time" point of view, the F_{ab} of Eqn. (36) represents a packet of electromagnetic waves which moves through space. (The question we must answer, from Maxwell's equations, is is "How does it move.") The φ in (36) is just a number. We are interested in (36) in the limit of large α. When α is large, the slightest change in φ causes $\alpha\varphi$ to go through many multiples of 2π. Hence, $\sin(\alpha\varphi)$ oscillates between $+1$ and -1 many times. That is the limit of (36) for large α is the limit of high frequency and short wavelength. i.e., the optical limit.

We ask: what are the conditions of φ and $\overline{F}_{ab}$ in order that (36) satisfy Maxwell's equations, (25) and (26), in the limit of large α? (We set $J^a = 0$, for we are interested in free wave solutions.) Substituting (36) into (25) and (26), we obtain

$$\sin(\alpha\varphi)\nabla^{[a}\overline{F}^{bc]} + \alpha\cos(\alpha\varphi)(\nabla^{[a}\varphi)\overline{F}^{bc]} = 0$$
$$\sin(\alpha\varphi)\nabla_b\overline{F}^{ab} + \alpha\cos(\alpha\varphi)(\nabla_b\varphi)\overline{F}^{ab} = 0,$$

respectively. For α large, the first term in each equation above is negligible compared to the second, Hence, setting $k^a = \nabla^a\varphi$, we have a solution in the optical limit provided

$$k^{[a}\overline{F}^{bc]} = 0 \qquad\qquad (37)$$

$$k_b\overline{F}^{ab} = 0 \qquad\qquad (38)$$

Contracting (37) with k_a, using (38), we obtain $\frac{1}{3}k_ak^a\,\overline{F}^{bc} = 0$. Hence, $k_ak^a = 0$, i.e., k^a is a null vector field. But now $k^b\nabla_bk_a = k^b\nabla_b\nabla_a\varphi = k^b\nabla_a\nabla_b\varphi = k^b\nabla_ak_b = \frac{1}{2}\nabla_a(k^bk_b) = 0$. Hence, k^b is tangent to null geodesics. All that remains is to interpret these results.

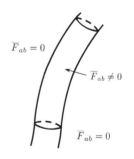

First note that the surfaces of constant phase for the wave are the surface of constant φ. Hence, $k^a = \nabla^a\varphi$, which is normal to these surfaces of constant phase, is the direction of propagation of the wave in space-time. We conclude that *light rays are null geodesics in space-time.* (More precisely, a packet of electromagnetic waves, in the optical limit, moves on a null geodesic.) Thus, light goes, not along any old null curve, but along a particular type of null curve. (This is ok. If you want to send out a pulse of light with a flashlight, you get to choose the event at which the light is sent, and the direction (in space) in which the light is emitted. Given a point in space-time, a four-velocity there, and a direction orthogonal to that four-velocity, there is precisely one null direction whose projection orthogonal to the four-velocity is the given (spatial) direction. But a null direction at a point of space-time uniquely determines a corresponding null geodesic.)

A direct physical interpretation of k^a can be obtained by introducing an observer. Set $\alpha = 1$ in (36). (That is, choose α large

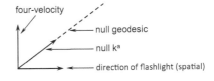

enough so the optical limit is valid, then rescale φ by a constant factor to get $\alpha = 1$. Let our observer have four-velocity ξ^a. Now, $\xi^a k_a = \xi^a \nabla_a \varphi$ is the rate of change (with time) of the phase of the wave as seen by our observer, that is to say, $\omega = -\xi^a k_a$ is the apparent (angular) frequency of the wave, as seen by our observer. Set $K_a h^a{}_b k_b$, a spatial vector according to our observer. Then if v^a is a unit vector in the direction of K^a, $v^a K_a = v^a \nabla_a \varphi$ is the rate of change of the phase of the wave with respect to distance, in the spatial direction v^a (according to this observer). When the phase has gone through 2π, then one has passed form one crest of the wave to the next. Hence, $\lambda = 2\pi / v^a K_a$ is the wavelength of the light, as seen by this observer. This can be simplified by recalling that v^a is a unit vector in the direction of K_a. Hence,

$$\lambda = 2\pi / v^a K_a = 2\pi \left[\frac{K^a}{(K^m K_m)^{1/2}} \right] K_a = 2\pi (K^m K_m)^{-1/2}$$

$$= 2\pi (h^{ab} k_a k_b)^{-1/2} = 2\pi [(g^{ab} + \xi^a \xi^b) k_a k_b]^{-1/2} == -2\pi (\xi^a k_a)^{-1}$$

Note that the angular frequency and wavelength are related by $\lambda = 2\pi \omega^{-1}$, as we would expect for a wave traveling (according to our observer) at the speed of light.

All the usual formulae for Doppler shifts, aberration, etc. follow immediately and easily from the remarks above. As an example, we work out the formula for the Doppler effect. Let two observers pass through the event p in space-time, and let

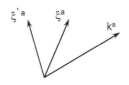

their four-velocities be ξ^a and ξ'^a. A light ray (with its k^a) also comes to the event p. The two observers see it, and assign to it frequencies $\omega = -\xi^a k_a$ and $\omega' = -\xi'^a k_a$. How are these frequencies related? It depends, of course, on the relationship between ξ^a, ξ'^a and k_a. Let us suppose that these three vectors are linearly dependent. (Physical interpretation: each observer sees the other moving in the direction, or opposite the direction, of the light ray.) Thus, we are doing what is called the longitudinal Doppler shift. (For the transverse Doppler shift, one would require that the projection of ξ' orthogonal to ξ is orthogonal to the projection of k^a orthogonal to ξ. That is, one would require that the ξ-observer see the light ray and the ξ'-observer moving in orthogonal directions.) Since ξ^a, ξ'^a and k^a are dependent, k^a is a linear combination of ξ^a, ξ'^a : $k^a = m\xi^a + m'\xi'^a$. Setting $k^a k_a = 0$, and solving for m and m' (we actually care about their ratio), we have $k^a = s[-\xi^a(\xi^m\xi'_m - \sqrt{(\xi^m\xi'_m)^2 - 1}) - \xi'^a]$ where s is number. Hence,

$$\omega = -\xi_a k^a = s\sqrt{(\xi^m\xi'_m)^2 - 1}$$
$$\omega' = -\xi'_a k^a = s[(\xi^m\xi'_m)^2 - 1 - (\xi^m\xi'_m)\sqrt{(\xi^m\xi'_m)^2 - 1}]$$

So,

$$\frac{\omega'}{\omega} = -\xi^m\xi'_m + \sqrt{(\xi^m\xi'_m)^2 - 1}$$

This is the required formula. It is convenient, however, to express it in terms of apparent relative speed between the observers. (See Eqn. (10), writing ξ'^a for ξ^a.) Substituting (10), we obtain, finally,

$$\frac{\omega'}{\omega} = \sqrt{\frac{1+v}{1-v}}$$

this will be recognized as the familiar formula for the longitudinal Doppler shift.

As far as I am aware, Lorentz transformations are never a good way to obtain the standard formulae of special relativity. Instead, one can set up the problem in terms of the relevant vectors and tensors, work out the inner product which yield what one wants, and solve for what one wants in terms of what one is given. Furthermore, these formulae are, in my opinion, almost never useful. The tensors are simple and clear. The formulae which relate these tensors to the experience of an observer should, perhaps, be seen once, to make it clear how they can be obtained. Beyond that, one can happily live with just the tensors.

18. Geodesic Deviation

Let γ be a straight line in Euclidean 3-space, with unit tangent vector ξ^a. Let γ' be a nearby straight line. At each point of γ, let η^a be the connecting vector from that point, orthogonal to γ, to the corresponding point of γ'. It should be clear geometrically that $\xi^n \xi^n \nabla_n \nabla_n \eta^a = 0$, i.e., the second derivative of the connecting vector along γ vanishes. Now suppose we replace Euclidean 3-place by space-time, and the straight line by geodesics. Then the second derivative of the connecting vector along the geodesic is,

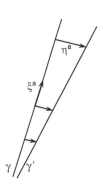

rather then zero, an expression involving the curvature of space-time. We now derive this expression. It gives a pretty, intuitive, "geometrical" picture of what curvature in space-time does.

It is convenient to replace the intuitive discussion involving "nearby curves" above by a one-parameter family of curves. We thus begin as follows. Let $S = R^2$ (the plane) with coordinates u, v. Introduce vector fields u^a and v^a on S such that $u^a \nabla_a u = v^a \nabla_a v = 1$, $u^a \nabla_a v = v^a \nabla_a u = 0$. (It is not difficult to

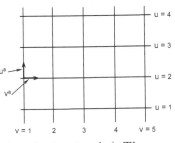

check that these properties determine u^a and v^a uniquely.) Thus, we have a two-dimensional "grid".

The idea is to "insert" this grid into space-time, to obtain our one-parameter family of curves (v will label the various curves, and u will be a parameter along each curve). This we do as follows: let $\psi : S \to M$ be a smooth mapping. Then, fixing v, $\psi(u, v)$, as u varies, describes a curve in space-time, where u serves as a parameter along this curve. Thus, we obtain a one-parameter family of curves (labeled by v). Denote by ξ^a and η^a, respectively, the images of u^a and v^a under this mapping. That is, set $\xi^a = \vec{\psi}\, u^a$, $\eta^a = \vec{\psi}\, v^a$. Evidently, ξ^a is just the tangent vectors to our curves, while η^a is the connecting vector between successive curves. Thus, by introducing a family of curves, and taking the "derivative of position in space-time with respect to the curve (i.e., with respect to this parameter)", we obtain, in a clean way a vector η^a which represents, intuitively, the "connecting vector between curves".

We are interested, not in arbitrary one-parameter families of
curves, but rather in one-parameter
families of timelike geodesics. Hence,
we set $\xi^a \xi_a = -1$, and $\xi\ -^b \nabla_b \xi^a =$
0. Thus, u measures proper time
along each timelike geodesics. Furthermore, it fellows from our con-
struction that $\xi_m \nabla_m \eta^a = \eta^m \nabla_m \xi^a$.
(Proof: This says that the Lie
derivatives of η in the ξ-direction
vanishes. But, clearly, the Lie

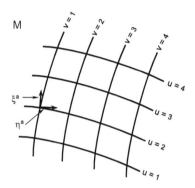

derivative of v^a in the u-direction vanishes. The result now follows from the fact that Lie derivatives are carried forward by smooth map-pings.) Geometrically, $\xi_m \nabla_m \eta^a = \eta^m \nabla_m \xi^a$ states that, if the "foot" of the infinitesimal displacement sits on one geodesic, and the "tip of the arrow" sits on another, then, as η^a is introduced at succes-sive points along the geodesics, it continues to point from the same geodesic. In other words, this is precisely the condition that η^a really represent the "connecting vector" from one geodesics to a neighboring one.

Having set up what is to be calculated, the calculation itself is easy:

$$\xi^m \xi^n \nabla_m \nabla_n \eta_a = \xi^m \nabla_m (\xi^n \nabla_n \eta_a) - (\xi^m \nabla_m \xi^n) \nabla_n \eta_a$$
$$= \xi^m \nabla_m (\xi^n \nabla_n \eta_a)$$
$$= \xi^m \nabla_m (\eta^n \nabla_n \xi_a)$$
$$= (\xi^m \nabla_m \eta^n) \nabla_n \xi_a + \xi^m \eta^n \nabla_m \nabla_n \xi_a$$
$$= (\xi^m \nabla_m \eta^n) \nabla_n \xi_a + \xi^m \eta^n \nabla_n \nabla_m \xi_a + \xi^m \eta^n R_{mnac}\, \xi^c$$
$$= (\xi^m \nabla_m \eta^n) \nabla_n \xi_a + \eta^n \nabla_n (\xi^m \nabla_m \xi_a) - (\eta^n \nabla_n \xi^m) \nabla_m \xi_a$$
$$+ \xi^m \eta^n R_{mnac}\, \xi^c$$
$$= -(R_{manb}\, \xi^m \xi^n) \eta^b$$

where, in the first step, we have differentiated by parts; in the second, we have
used $\xi^m \nabla_m \xi^n = 0$; in the third, we have
used $\xi^n \nabla_n \eta^a = \eta^n \nabla_n \xi^a$; in the fourth, we
have expanded using the Leibniz rule: in
the fifth, we have used the definition of the
Riemann tensor, $\nabla_m \nabla_n \xi_a = \nabla_n \nabla_m \xi_a + R_{mnac} \xi^c$; in the sixth, we have differentiated the middle term in the preceding ex-
pression by parts. Finally, in the last step,
we have used the fact that the first and
third terms in the preceding expression cancel, and that the second
term is zero by geodesic-ness. We have:

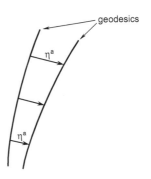

$$\xi^m \xi^n \nabla_m \nabla_n \eta_a = -(R_{manb}\, \xi^m \xi^n)\eta^b \tag{39}$$

This is called the *equation of geodesic deviation*. Note that, for flat
space ($R_{abcd} = 0$), we recover the observation at the beginning of this
section.

Why do we wind up with an equation
such as (39)? First note that curvature
could not affect a single geodesic, for there
would be nothing to compare it with (i.e.,
that there is no natural "thing which is like
a geodesic, but uninfluenced by a curvature") In light of this, curvature does the
best it can: it manifests itself when we
compare one geodesic with a nearby one.
Example: geodesic (great circles) on a 2-
sphere. The curvature "draws nearby geodesics together". Why is

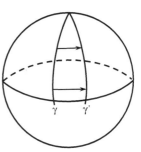

it that the second derivative of η^a (along the geodesic) appears on
the left in (39), rather than, say, the first derivative? The point here
is that $\xi^m \nabla_m \eta^a$ is freely at our disposal. We can "start out" our
two nearby geodesic running parallel to each other ($\xi^m \nabla_m \eta^a = 0$ initially), separating from each other ($\xi^m \nabla_m \eta^a \propto \eta^a$ initially)., moving
"sideways relative to each other" ($\xi^m \nabla_m \eta^a$ orthogonal to η^a initially).
However, after we fix how the two nearby geodesics begin, the geodesic
equations take over. In particular, they determine the second derivative of the connecting vector along the geodesic. Hence, the second
derivative is the first derivative not at our disposal, and hence the
first derivative, which possibly appear in (39). (Expressions for, e.g.,
$\xi^m \xi^n \xi^p \nabla_m \nabla_n \nabla_p \eta^a$ are obtained immediately from (39).)

Setting $K_{ab} = -R_{manb}\, \xi^m \xi^n$, Eqn. (39) becomes $\xi^m \xi^n \nabla_m \nabla_n \eta_a = K_{ab}\eta^b$. Note that K_{ab} is a symmetric, spatial tensor according to the observer with four-velocity ξ^a. ($K_{ab} = K_{(ab)}$. Proof: $R_{manb} = R_{nbma}$; $\xi^b K_{ab} = 0$. Proof: Since $R_{manb} = R_{ma[nb]}$, $\xi^m \xi^n \xi^b R_{manb} = 0$.) Hence, if η^a and $\xi^m \nabla_m \eta^a$ are initially orthogonal to ξ^m, this remains true along the geodesic. We can, without loss of generality, impose this orthogonality; and do so. There follows an intuitive discussion of the geometry of this K_{ab}.

Let, α, λ^b be an eigenvalue-eigenvector of K_{ab}, i.e., let $K_{ab}\lambda^b = \alpha\lambda_a$. Choose $\eta^a = \lambda^a$, and $\xi^m \nabla_m \eta^a = 0$ initially, where we write a dot for $\xi^m \nabla_m$. Then (39) gives $\ddot{\eta}^a = \alpha\eta^a$. In other words, this particular η^a merely lengthens or shortens (α positive or negative, respectively) as we move along the geodesic. Now consider the three independent eigenvalues-eigenvectors of K_{ab}. Each lengthens or shortens, depending on its eigenvalue. But a general η^a can be written as a linear combination

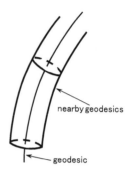

nearby geodesics

geodesic

of these eigenvectors. Thus, we have a complete description of the action of (39). We set tings up as follows. At some point on our basic geodesic, begin nearby geodesics all equidistant from the given geodesic, and running parallel to it (i.e., all have $\xi^m \nabla_m \eta^a = 0$) Then, to second order as we move away from this point along our basic geodesic (second order cause in (39)), the small 2-sphere described by the neighboring geodesics begins distorting into an ellipsoid. The axes of the ellipsoid are the eigenvectors of K_{ab}. Thus, if one eigenvalue is positive and two negative, we obtain a "cigar", while one negative and two positive yield a "pancake". Finally, note that the rate of change of the volume of the sphere is the sum of the eigenvalues, i.e., $K^m{}_m$.

We now repeat this description physically. You, the observer, are in free fall (i.e., a geodesic). At some instant, you release a lot of particles, all at rest relative to you, and all lying in a sphere of radius one foot from you. The derivative of the shape of the sphere (with time) is zero, since the particles were initially at rest. However, the second derivative with time results in a distortion of the sphere into an ellipsoid. The axes of the ellipsoid are the eigenvectors of K_{ab}. The second time rate of change of the volume of the sphere (as it goes into an ellipsoid) is $K^m{}_m$.

Clearly, the Riemann tensor is determined completely by its effects in causing geodesic deviation, i.e., by Eqn. (39).

19. Einstein's Equation

We shall now write down Einstein's equation – the equation analogous to $\nabla^2 \varphi = 4\pi G\rho$ in Newtonian theory.

We begin by asking the question: what is the relative behavior of nearby particles in Newtonian theory, i.e., what is "Newtonian geodesic deviation" like? Recall that the motion of a particle in Newtonian theory is given by $x^a(t)$, the position of the particle in Euclidean space as a function of time t. The equation of motion of such a particle (if uninfluenced by other forces) is $\ddot{x}^a = -\nabla^a \varphi$. Now suppose we have two nearby free particles, with positions x^a and $x^a + \delta x^a$. Then, to first order in δx^a, we have

$$(\delta x^a)^{\cdot\cdot} = (x^a + \delta x^a)^{\cdot\cdot} - (x^a)^{\cdot\cdot} = -\nabla^a \varphi(x + \delta x) + \nabla^a \varphi(x)$$

$$= -(\nabla^a \nabla^b \varphi)\delta x_b. \tag{40}$$

Eqn. (40) gives the effect of ordinary Newtonian gravitation on nearby particles. Suppose, for example, that we begin with a swarm of particles initially at rest with respect to the Earth, and above the Earth, where these particles form a 2-sphere. The particles are released. Since the Earth attracts the particles closer to it more than those farther away, a bit later the sphere becomes an ellipsoid, with the axis pointing toward the Earth. If, on the other hand, we began with particles forming a sphere about the center of the Earth, then a bit later they would still describe a sphere, but a smaller one (all particles fall toward the center of the Earth).

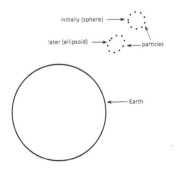

We regard Eqn. (40) as closely
analogous to Eqn. (39). Here, x^a
corresponds to η^a a dot corresponds
to $\xi^m \nabla_m$ ("time=derivatives"), whence
$\nabla_a \nabla_b \varphi$ is analogous to $R_{manb} \xi^m \xi^n$.
Thus, *because of the geodesic behav-*
ior of free particles in space-time,

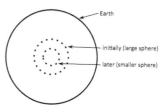

the quantity $R_{manb} \xi^m \xi^n$ *has the same effect on nearby free particles*
in space-time as has $\nabla_a \nabla_b \varphi$ *on nearby free particles in Newtonian*
theory.

The interpretation of $\nabla^2 \varphi = 4\pi G\rho$ in Newtonian theory should
now be clear. It says that the second time derivative of the volume of
a small sphere of free particles is given by the density of matter within
the sphere. (Note that $\nabla^2 \varphi$ is the sum of eigenvalues of $\nabla_a \nabla_b \varphi$.)
Thus, in the figure on the preceding page, there is no matter inside
the sphere, of particles. Hence, the particles begin to form an ellipsoid,
but with the same volume as the original sphere. In the figure on this
page, the decrease in the volume of the sphere with time is due to the
presence of matter (the Earth) inside the sphere of particles. Fix a
particle of mass m in Newtonian theory. Consider particles, begun at
rest, one foot from this one. Then $\nabla^2 \varphi = 4\pi G\rho$ states that, on the
average (over directions in which the second particle lies from the first)
the two particles fall closer together by Newton's law of gravitation.
Because of the presence of other masses outside this system, some
of the particles (on the one-foot sphere) may fall toward the central
particle more quickly; others more slowly. But $\nabla^2 \varphi = 4\pi G\rho$ fixes the
average.

Since $R_{manb} \xi^m \xi^n$ is analogous to the Newtonian $\nabla_a \nabla_b \varphi$, and since
we have $\nabla^2 \varphi = 4\pi G\rho$ in Newtonian theory, one would naturally set
$R_{ma} \xi^m \xi^n = 4\pi G\rho$ as the equation for the creation of curvature by
matter in general relativity. This will ensure that each free observer
sees nearby free particles falling in toward him, on the average, by an
amount determined by his ρ. In other words, this will ensure that our
observer, locally, sees the validity of Newtonian gravitation.

Unfortunately, $R_{ma} \xi^m \xi^n = 4\pi G\rho$ still contains this quantity ρ, the
Newtonian mass density. By what should it be replaced is space-time?
Unfortunately, there are two choices. One could set $\rho = T_{mn} \xi^m \xi^n$, the
local mass density seen by an observer with four-velocity ξ^m, where
T_{mn} is the total stress-energy. There is a second expression, based
on the fact that, in the Newtonian limit the mass density dominates
(i.e., pressures, momentum densities, etc. are all small). That is, in
the Newtonian limit, $T_{mn} = \rho \xi_m \xi_n$. Hence, we could equally well set
$\rho = -T$ as the space-time object to replace the Newtonian ρ.

More generally, we could consider some linear combination $-\rho = rT_{mn}\,\xi^m\xi^n - (1-r)T$, where r is a number, Then our generalization of $\nabla^2\varphi = 4\pi G\rho$ would be

$$R_{mn}\,\xi^m\xi^n = 4\pi G[rT_{mn} + (1-r)Tg_{mn}]\xi^m\xi^n. \tag{41}$$

Since this must hold for all unit timelike ξ^m, we have Einstein's equation:

$$R_{ab} = 4\pi G[rT_{ab} + (1-r)Tg_{ab}]. \tag{42}$$

In short, we obtain a one-parameter family (labeled by r) of generalization of the Newtonian law.

What value shall we assign to r? We first rewrite Eqn. (42). Contracting over "a" and "b", solving for T in terms of R, and substituting into (42), we have

$$R_{ab} - \frac{r-1}{3r-4}Rg_{ab} = 4\pi G\,r\,T_{ab} \tag{43}$$

Recall that the total stress-energy is conserved: $\nabla_b T^{ab} = 0$. Furthermore, Bianchi's identity yields $\nabla_m(R^{ab} - \frac{1}{2}Rg^{ab}) = 0$. The natural choice for r is that wish makes these equations consequences of each other. That is, we are led to set $r = 2$, so both sides of (43) are divergence-free, Thus, we obtain *Einstein's equation*:

$$R_{ab} - \frac{1}{2}Rg_{ab} = 8\pi\,G\,T_{ab} \tag{44}$$

(That both sides of (44) be conserved in actually required, as we shall see later, for the existence of an initial-value formulation of the theory.)

We regard, the Einstein equation, (44), as requiring that the curvature of space-time be so related to the matter distribution that free particles, responding to the curvature, behave as do free particles responding to the Newtonian potential behave in Newtonian theory. To put matters another way, it is clear that everything we have done has the correct special relativity limit (when $R_{abcd} = 0$). Eqn. (44) ensures that we also have the correct limit of Newtonian gravitation.

Note: $G = 6.67 \times 10^{-8}$ dyne cm^2/g^2 = 2.2×10^{-39} sec/g.

20. Einstein's Equation: General Remarks

Consider Maxwell's Equations, Eqns. (25) and (26). It is often the case in practice that one adopts the following attitude toward these equations. The charge-current J^a is given, and one wishes to solve (25) and (26) for the electromagnetic field, F_{ab}. This would be the situation, for example, if we were given the current in a system of wires (i.e., given J^a), and wished to find the resulting the electromagnetic field. In other situations, however, the charge-current is strongly influenced by F_{ab}, and so no such procedure is available for solving (25) and (26). This would be the case, for example if we were dealing with a cloud of charged particles, so the value of F_{ab} determines the motion of the cloud, and hence J^a. Similar remarks apply to $\nabla^2 \varphi = 4\pi G \rho$ in Newtonian gravitation. To find the gravitational field of a given object (i.e., an elephant), one solves for φ, with ρ given. To determine the motion of the planets, φ affects what ρ is, so no such simple procedure is available.

How does Einstein's equation fit into this framework? *It is unreasonable physically to regard T_{ab} as given* (on a manifold), *with Einstein's equation (44) to be solved for the metric g_{ab}.* The reason is that T_{ab} itself is a quantity which refers, not only to "matter", but also to "geometry". Consider, for example, the stress-energy of a fluid, $T_{ab} = (\rho + p)\xi_a \xi_b + p g_{ab}$. Now, ξ^a, the four-velocity of the fluid, is unit ($\xi^a \xi^b g_{ab} = -1$); ρ is the mass density in grams/sec³; the metric occurs in the second term. Everywhere, we see the metric, directly or indirectly, in the stress-energy. One should not regard "specifying T_{ab}" as the same as "specifying what the matter is doing". The relation between T_{ab} as a tensor field and "what the matter is doing" involves the metric. It appears that it is simply impossible to make any reasonable description of matter without the notions of space and time provided by the metric.

In light of the remarks above, one might be tempted to proceed

in the other direction. We give g_{ab} and "solve" (trivially) (44) for
T_{ab}. One could indeed regard the result as a "solution" of Einstein's
equation. The difficulty is that the resulting T_{ab} will not, without
extraordinary luck, resemble any kind of matter of interest.

As far as I am aware, there exists no technique for generating all
physically interesting solutions of Einstein's equation. The collection
of known exact solution is relatively small, and this makes every exact solution potentially valuable (either because it might provide a
prediction of general relativity, or – often as important – because it
might provide insight into the structure of the theory itself). Those
solutions which are available are obtained by special techniques, e.g.,
assumptions of symmetry, or assumptions of simple structure for the
Riemann tensor. We shall later introduce some techniques for obtaining solutions of Einstein's equation.

Recall the discussion of Sect. 16. We can regard gravitation as just
another "type of matter". The corresponding tensor field is the metric,
g_{ab}. The equation on this field is Einstein's equation. Of course, the
metric needn't have a stress-energy: that's the mechanism by which
other types of matter influence the metric. The situation is similar to
that of electromagnetism. All "types of charge" end up producing a
charge-current density, J^a. If several "types of charge" are available,
we add their contributions to obtain the total charge-current J^a. This
acts as a source for the electromagnetic field. This total charge-current
must have vanishing divergence (in order that there exist solutions of
(25), and (26)) If in fact the "types of charge" are able to exchange
charge between them, then the individual J^a's for the individual types
of charge will not be conserved: only the total. The electromagnetic
field is not obligated to come up with a charge-current for it. (The
analogy: J^a for T_{ab}; vanishing divergence for conserved; F_{ab} for g_{ab}.)

What is so special, from the point of view above, that sets gravitation apart from other fields? It is the universality of gravitation.
The metric enters into everybody else's stress-energy – it effects all
types of matter. Every type of matter produces a stress-energy, and
hence affects the geometry of space-time. (In the electromagnetic case,
by contrast, some things are uncharged, and so do not interact with
electromagnetism.) Hence, all possible measuring instruments are affected by g_{ab}. Because of this universality, any attempt to measure
features of the geometry of space-time will be influenced by gravitation. Hence, gravitation gets inseparably intertwined with geometry.
Thus, the metric, g_{ab} has two roles: as giving geometry of space-time,
and as a gravitational potential. (Think of g_{ab} as analogous to Newtonian φ. Then $\nabla_a \nabla_b \varphi$ is analogous to the "second derivative" of the
metric, i.e., to Riemann tensor. That's what we want from Sect. 19.)

21. The Friedmann Solutions

The information presently available about the structure of our Universe is rather meagre. However, the following two features do appear to be comparatively well-established. *The dominant matter in the Universe* (with regard to the dynamics of the Universe) *is the matter in the galaxies, which may be regarded as dust.* Essentially all the matter of which we are aware in the Universe is that in the galaxies. There are some 10^{11} galaxies, enough so that we might expect to be able to regard each as a "particle of dust". There are, of course, other things in the Universe (e.g., electromagnetic radiation), but these contribute to the total stress-energy a negligible amount. *The Universe is spatially isotropic, as seen from any galaxy.* (The mathematical formulation of "spatially isotropic" will come later.) First note that the Universe is not isotropic, in its details as seen from Earth. For example, we see nearby galaxies in some directions and not in others. But these are local features. In the large (i.e., over distances large compared with the distance between galaxies), the Universe does indeed appear isotropic from the Earth. For example, we see about as many galaxies in one region of the sky as in another, on the average. From bitter previous experience, we do not wish to regard the location of the Earth in the Universe as somehow special. Hence, we are led to assume isotropy for everybody.

It is an easy matter to obtain the most general solution of Einstein's equation compatible with the features above. If general relativity is correct, and if the features above are indeed valid for our Universe, then one of these solutions will describe the overall behavior of our Universe. We now obtain these, the Friedmann solutions.

The matter is dust. Hence, we have the density ρ of the dust, the four-velocity ξ^a of the dust, and

$$T_{ab} = \rho \xi_a \xi_b \tag{45}$$

Next, consider $\nabla_a \xi_b$. The antisymmetric part, $\nabla_{[a} \xi_{b]}$ is an antisymetric tensor field in space-time. Just as we did for the electromagnetic field in Sect.13. this $\nabla_{[a}\xi_{b]}$ can be decomposed (by observer with four-velocity ξ^a, i.e., by us) into two spatial vectors. But this would violate isotropy! If a spatial vector could be obtained from the velocity field of the galaxies and the metric, then some directions in space would be preferred over others. Since we wish to impose isotropy – since we wish that no spatial directions are preferred – we must have $\nabla_{[a}\xi_{b]} = 0$. Thus, $\nabla_a\xi_b$ must be symmetric in "a" and "b". Next, note that $\xi_a^{\nabla}\xi_b = 0$ (since ξ^b is unit), and $\xi_a\nabla_a\xi_b = 0$ (geodesic motion of dust particles). Hence, $\nabla_a\xi_b$ is a (symmetric) spatial tensor (according to observer with four-velocity ξ^a). Consider any eigenvector λ^a of $\nabla_a\xi_b$. That is, let λ^a be spatial, with $\lambda^b(\nabla_a\xi_b) = \alpha\lambda_a$. This spatial vector would be preferred (violating isotropy) unless $\mu^b(\nabla_a\xi_b) = \alpha\mu_a$ for all spatial μ_b. Thus,

$$\nabla_a\xi_b = \alpha h_{ab} \qquad (46)$$

for some scalar fields α, where, as usual, we set $h_{ab} = g_{ab} + \xi_a\xi_b$. To interpret this α physically, contract (46): $\nabla^a\xi_a 3\alpha$. That is, the divergence of ξ^a is 3α, so α represents the "rate of expansion of the Universe". We shall call α the *Hubble field* (for reasons which will emerge shortly).

To impose Einstein's equation, we need the Ricci tensor, which is obtained from the Riemann tensor, which is obtained by commuting derivatives. The obvious thing on which to commute derivatives is ξ_c. We have

$$\begin{aligned}
\nabla_a(\nabla_b\xi_c) &= (\nabla_a\alpha)h_{bc} + \alpha\nabla_a h_{bc} \\
&= \dot\alpha\xi_a h_{bc} + \alpha\nabla_a h_{bc} = -\dot\alpha\xi_a h_{bc} + \alpha\nabla_a(g_{bc} + \xi_b\xi_c) \\
&= -\dot\alpha\xi_a h_{bc} + 2\alpha(\nabla_a\xi_{(b)}\xi_{c)} \\
&= -\dot\alpha\xi_a h_{bc} + 2\alpha^2 h_{a(b}\xi_{c)}
\end{aligned} \qquad (47)$$

where, for the first step, we have substitute (46); for the second, used $\nabla\alpha = -\dot\alpha\xi_a$, where a dot denotes $\xi^m\nabla_m$ (by isotropy, $\nabla_a\alpha$ cannot have a non vanishing component orthogonal to ξ_a. Hence, $\nabla_a\alpha$ a multiple of ξ_a; for the third, used $h_{bc} = g_{bc} + \xi_b\xi_c$; for the fourth, expanded using the Leibniz rule; and, for the fifth, again used (46). Now antisymmetrize (47) over "a" and "b", using $\nabla_{[a}\nabla_{b]}\xi_c = \frac{1}{2}R_{abcd}\,\xi^d$:

$$R_{abcd}\,\xi^d = -2(\dot\alpha + \alpha^2)\xi_{[a}h_{b]c} \qquad (48)$$

Thus, we obtain immediately from (46) a simple expression for "most" of the Riemann tensor. Isotropy is a very strong condition, which produces a considerable simplification.

To use Einstein's equation, we need the Ricci tensor. Hence, we contract (48) over "a" and "c" (after raising one of the indices, of course. We won't normally state that explicitly.) Hence, from (48):

$$R_{bd}\,\xi^d = -3(\dot{\alpha} + \alpha^2)\xi_b \tag{49}$$

We next use Einstein's equation, (44) but in the form $R_{ab} = 8\pi\,G(T_{ab} - \frac{1}{2}T g_{ab})$, where $T = T^m{}_m$. (To obtain this form, contract (44) to obtain $R = -4\pi\,G T$. Substitute this for R in (44).) Hence, from (45),

$$R_{ab} = 8\pi\,G(\rho\xi_a\xi_b + \frac{1}{2}\rho g_{ab}) \tag{50}$$

Finally, substitute R_{ab} from (50) into (49) to obtain

$$\dot{\alpha} + \alpha^2 = -\frac{4}{3}\pi\,G\,\rho \tag{51}$$

Eqn. (51) has a simple physical interpretation. The $\dot{\alpha}$ is "the rate of charge with time of the rate of expansion of the Universe". Hence, $\dot{\alpha}$ is an "acceleration". Ignoring factors and the $\dot{\alpha}$ terms, Eqn. (51) says that "the Universe has negative acceleration, with magnitude ρ". In other words, "the attraction of all those galaxies for each other tends to slow the rate of expansion of the Universe". This is (part of) the effect of the dust on the geometry of space-time required by Einstein's equation.

Eqn. (51) alone cannot be solved. We have $\alpha(t)$ and $\rho(t)$ (where t is proper time along a galaxy world-line). (Clearly, we expect the mass density of galaxies, ρ, to change with time as the universe expands.) To obtain a second equation, we use conservation of dust: $0 = \nabla_a(\rho\xi^a) = \rho\nabla_a\xi^a + \xi^a\nabla_a\rho$. Substituting (46),

$$\dot{\rho} + 3\alpha\rho = 0 \tag{52}$$

Eqn. (52) is also clear physically. It states that "the mass density of the dust decreases at a rate proportional to the rate of expansion of the Universe" – exactly what we would expect from conservation of dust particles.

We now wish to solve the simultaneous Eqns. (51) and (52). Since α is the "rate of expansion of the Universe", it is convenient to introduce a function $r(t)$ such that α is the "rate of expansion of r". Set $\alpha = \dot{r}/r$. (Remark on the physical interpretation of r: Let η^a be a spatial vector along a galaxy world-line, and suppose η^a joins this world-line to the world-line of nearby galaxy.

Then, setting $r = (\eta^a\eta_a)^{\frac{1}{2}}$, we have $\dot{r}/r = \alpha$. (Hint: Use $\mathscr{L}_\xi\eta^a = 0$.) Thus, r could be the "spatial distance of the Andromeda from us".) Setting $\alpha = \dot{r}/r$ in (52), we have $3\dot{r}\rho + r\dot{\rho} = 0$, or $(r^3\rho) = $ const. (What does $r^3\rho$ const. mean physically?) We make the following choice of this constant (the constant is at our disposal because $r(t)$ is only defined up to a constant factor):

$$\frac{4}{3}\pi G \rho = r^{-3} \tag{53}$$

Substituting (53) and $\alpha = \dot{r}/r$ into (51), we easily obtain

$$\ddot{r} = -r^{-2} \tag{54}$$

Eqn. (54) also has a simple physical interpretation. Think of $r(t)$ as a "characteristic size of the Universe". Then $\ddot{r}$ is the "acceleration". Hence, from (54), we regard $-r^{-2}$ as an "effective (attractive, inverse r-squared) force". In fact, Eqn. (54) is precisely the equation describing a brick thrown radially from the Earth (where $r = $ distance of brick from Earth's center). The acceleration of the brick ($\ddot{r}$) equals the force on the brick ($-r^{-2}$). It is now clear, from our intuitive understanding of bricks, what the solutions of (54) (and, hence, of (51) and (52)) will look like. Bricks which are thrown upward with a small initial speed soon fall back. Those which have a large initial speed escape from the Earth. There is a critical speed between these extremes. These bricks escape, but with a speed which diminishes toward zero in the limit as the brick gets far away from the Earth.

The analogy above also suggests how to solve (54): use conservation of energy. Multiplying (54) by $\dot{r}$, we have $2r^{-1} - (\dot{r})^2 = c = $const. (For the brick, this is the binding energy. The brick returns precisely when c is positive.) Thus, $\dot{r} = \pm(2r^{-1} - c)^{\frac{1}{2}}$, which is easily solved with a table of integrals. It is simplest to express the solution in parametric form:

$$
\begin{array}{lll}
c > 0 & r = c^{-1}(c - \cos x) & t = c^{-3/2}(x - \sin x) \\[2mm]
c = 0 & r = \dfrac{1}{2}x^2 & t = \dfrac{1}{6}x^3 \\[2mm]
c < 0 & r = (-c)^{-1}(\cosh x - 1) & t = (-c)^{-3/2}(\sinh x - x)
\end{array} \tag{55}
$$

where x is a parameter. These functions are graphed below.

These graphs are, of course, exactly what we would expect from the brick analogy, Of course, given $r(t)$, one determines immediately $\alpha(t)$ and $\rho(t)$.

The discussion above refers to the "dynamics" of the Friedmann solutions. We now describe their "spatial

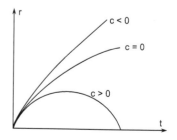

geometry". This discussion is based on the following remarks: if $W_{a...c}$ is any spatial tensor (according to ξ^a), then $\nabla_a W_{a...c}$, with all its indices projected orthogonal to ξ^a, is the spatial derivative of this tensor. (Proof: $\nabla_a h_{bc}$, with all its indices projected, gives zero. See definition of covariant derivative associated with a metric.) Let l_a be any spatial vector field in our space-time. Then

$$\frac{1}{2}\mathscr{R}_{abcd}\, l^d = h_{[a}{}^r h_{b]}{}^s h_c{}^t \nabla_r[h_s{}^m h_t{}^n \nabla_m l_n] \tag{56}$$

by definition of the spatial Riemann tensor, $\mathscr{R}_{abcd}$. Expanding the right side of (56) by the Leibniz rule,

$$\begin{aligned}\frac{1}{2}\mathscr{R}_{abcd}\, l^d &= h_{[a}{}^r h_{b]}{}^s h_c{}^t[(\nabla_r h_s{}^m)h_t{}^n \nabla_m l_n \\ &\quad + h_s{}^m(\nabla_r h_t{}^n)\nabla_m l_n + h_s{}^m h_t{}^n \nabla_r \nabla_m l_n]\end{aligned} \tag{57}$$

We consider the three terms on the right, one at a time. For the first term, $\nabla_r h_s{}^m = \nabla_r(\delta_s{}^m + \xi_s \xi^m) = \xi_s \nabla_r \xi^m + \xi^m \nabla_r \xi_s$. But ξ_s is annihilated by the $h_{b]}{}^s$ in in (57). Hence, this first term becomes $h_{[a}{}^r h_{b]}{}^s h_c{}^t[\xi^m(\nabla_r \xi_s)h_t{}^n \nabla_m l_n]$. But this vanishes, for $\nabla_r \xi_s$ is symmetric, while $h_{[a}{}^r h_{b]}{}^s$ is skew in "r" and "s". Thus, the first term on the right in (57) does not contribute. For the second term, we proceed similarly: $\nabla_r h_t{}^n = \xi_t \nabla_r \xi^n + \xi^n \nabla_r \xi_t$. The ξ_t is annihilated by $h_c{}^t$. Hence, using $\nabla_r \xi_t = \alpha h_{rt}$, this second term becomes $h_{[a}{}^r h_{b]}{}^s h_c{}^t[h_s{}^m \xi^n \alpha h_{rt} \nabla_m l_n]$. To simplify further, note that $\xi^n \nabla_m l_n = \nabla_m(\xi^n l_n) - l^n \nabla_m \xi_n = 0 - l^n \nabla_m m \xi_n = -\alpha l_m$, where we have again used (46). Thus, using $h_a{}^m h_{mn} = h_{an}$, the second term on the right in (57) becomes, finally, $-\alpha^2 h_{c[a} l_{b]}$. Mercifully, the third term on the right in (57) is easy. It is just $h_{[a}{}^r h_{b]}{}^s h_c{}^t \nabla_r \nabla_s l_t$, which, by the definition of the Riemann tensor, is just $\frac{1}{2}h_a{}^r h_b{}^s h_c{}^t R_{rstd}\, l^d$. Substituting these evaluations in the right side of (57), we obtain

$$\frac{1}{2}\mathscr{R}_{abcd}\, l^d = \frac{1}{2}h_a{}^m h_b{}^n h_c{}^p R_{mnpd}\, l^d - \alpha^2 h_{c[a} h_{b]d} l^d \tag{58}$$

But, since l^d is an arbitrary spatial vector, (58) implies

$$\mathscr{R}_{abcd} = h_a{}^m h_b{}^n h_c{}^p h_d{}^q R_{mnpq} - 2\alpha^2 h_{c[a} h_{b]d} \tag{59}$$

Eqn. (59) relates the spatial curvature, $\mathscr{R}_{abcd}$, to the curvature of space-time, R_{abcd}, and α. The idea is to use (59) to determine the spatial curvature.

Contracting (59) over "b" and "d", we have

$$\mathcal{R}_{ac} = h_a{}^m h_c{}^p h^{nq} R_{mnpq} - 2\alpha^2 h_{ac}$$
$$= h_a{}^m h_c{}^p R_{mp} + h_a{}^m h_c{}^p \xi^n \xi^q R_{mnpq} - 2\alpha^a h_{ac} \tag{60}$$

Now we are in great shape. The Ricci tensor of space-time, R_{mn}, appears in (60), but we have an expression for it from Einstein's equation, (50). From (50), $h_a{}^m h_c{}^n R_{mn} = 4\pi G \rho h_{ac}$. The expression $R_{amcn} \xi^m \xi^n$ appears in (60), but we have essentially evaluated that in (48). Contracting (48) with ξ^b, we obtain $\mathcal{R}_{ambn} \xi^m \xi^n = -(\dot\alpha + \alpha^2) h_{ab}$. Thus, (60) becomes

$$\mathcal{R}_{ac} = h_{ac}[4\pi G \rho - \dot\alpha - 3\alpha^2] \tag{61}$$

Thus, the spatial Ricci tensor is a multiple of the spatial metric. Such a space (when the Ricci tensor is a multiple of the metric) is said to be of *constant curvature*.

To simplify (61) still further, substitute (53) for ρ, and $\dot r/r$ for α. Then, using $c = 2r^{-1} - (r)^2$, we have

$$\mathcal{R}_{ac} = 2r^{-2} c\, h_{ac} \tag{62}$$

The sign of the scalar curvature of space depends on the sign of c (the "binding energy"). For c positive (the Universe recollapses), space is positively curved. It is natural that r^{-2} should appear on the right in (62). After all, r is a "characteristic size" of the Universe, and $\mathcal{R}_{ac}$ is curvature, units $\sec^{-2}$. (If you blow up a balloon, the curvature of the balloon goes like r^{-2}, where r is the radius of the balloon.)

22. The Friedmann Solutions: Continued

We begin with some remarks on the Friedmann solutions as a whole. We then discuss each of the three types individually.

The fundamental quantities in the Friedmann solutions are α (the expansion rate), ρ (the mass density of galaxies), and $\mathscr{R}$ $(= \mathscr{R}^m{}_m$, the scalar curvature of space). These three quantities are not independent of each other, but satisfy a single identity;

$$\mathscr{R} = 16\pi\, G\, \rho - 6\alpha^2 \tag{63}$$

(Proof: Contract (61), and use (51).) Thus, we may regard α and ρ as the independent quantities, with $\mathscr{R}$ expressed in terms of these by (63). If, at any point of space-time, the values of α and ρ are given, then the values of α and ρ for all t are thereby determined by (51) and (52). Thus, these two quantities would have to be measured at the present Epoch of our Universe to determine in which Friedmann solution we live, and in what portion of that solution we are.

The class of solutions obtained in Sect. (21) forms a one-parameter family. (It takes two measurements now to determine the structure of our Universe, but that gives us not only which solution we live in, but also where in the course of the evolution of that solution we are now.) To label this one-parameter family of solutions, note that Eqns. (51) and (52) admit a conserved quantity (i.e., a function of α and ρ which remains constant as α and ρ undergo (51) and (52).) This conserved quantity is

$$c = \left(\frac{4}{3}\pi\, G\, \rho\right)^{-2/3}\left[\frac{8}{3}\pi\, G\, \rho - \alpha^2\right] \tag{64}$$

(Proof: Eliminate r from $c = 2r^{-1} - (r)^2$.) As the solution evolves, the values of α and ρ change, but c remains a constant. Meanwhile, of course, $\mathscr{R}$ evolves via (63).

c negative. The Universe expands from an initial singularity ($\rho = \infty$) (the "big bang"), and continues expanding forever thereafter ($\alpha > 0$). The rate of expansion decreases monotonically with time (physically, caused by the attraction of the galaxies for each other). In the early stages of the expansion (near the singularity) the "size of the Universe", $r(t)$, goes like $r = \frac{1}{2}(6t)2/3$. (This early behavior is not very significant physically. At this epoch, we expect the matter to be very dense, highly compressed. Thus, the matter should exert pressure at this epoch, so a fluid would be a better approximation to the matter then dust.) In the limit of large t, $r = (-c)^{1/2}t$. Thus, the rate of expansion approaches a constant at large t.

Space has constant negative curvature, i.e., we have hyperbolic space. The radius of curvature of space (r) decreases monotonically during the expansion.

c zero. The Universe expands from an initial singularity, and continues expanding thereafter, but at an ever decreasing rate. The size of the Universe varies with t according to $r = \frac{1}{2}(6t)2/3$. Thus, the rate of expansion is infinitive at the early epoch, and decrease to zero in the limit of large t.

Space has constant zero curvature. That is to say, space is flat.

c positive. The Universe expands from an initial singularity, with $r = \frac{1}{2}(6t)2/3$ at the earliest epochs. The rate of expansion is infinite in the earliest epochs. This rate of expansion decrease, eventually passing through zero. Thus, the Universe stops expanding, and begins contracting. The maximum occurs at time $t = \pi c^{-3/2}$ from the initial singularity. The Universe recontracts, in the same amount of time, to a final singularity.

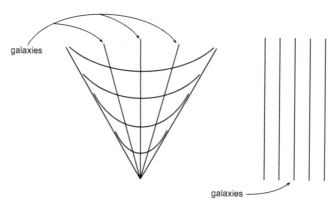

galaxies

galaxies ———

Space has constant positive curvature. Thus, the metric of space is that of 3-sphere. In particular, the Universe is closed spatially. The radius of curvature of space is zero at the singularities, but decrease to a maximum at the moment of maximum expansion. This maximum value is $\mathscr{R} = \frac{3}{2}c^3$. Suppose we set $\rho = 0$ in the Friedmann solutions? Then (52) is satisfied identically, and (51) has solutions $\alpha = t^{-1}$ and $\alpha = 0$. In the first case, $\mathscr{R} = -6t^{-2}$; in the second, $\mathscr{R} = 0$. Thus, in the first case, we have an expanding Universe of negative spatial curvature; in the second, a static Universe of zero spatial curvature. It is not difficult to check that these are both Minkowski space, but sliced up in different ways.

In the first case, the galaxies diverge from a common origin in Minkowski space. In the second case, the galaxies are parallel straight lines.

23. Friedmann Solutions: The Observations

We have seen that our Universe appears to have the following properties: the dominant form of matter is dust (galaxies), and space is isotropic about each event in space-time. The only solutions of Einstin's equation which possess these features are the Friedmann solutions. Thus, we would expect that one of these Friedmann solutions describes the large- scale behavior of our own Universe. Which one? What observations in our Universe can be compared with the predictions of the Friedmann solutions? We shall be concerned with these questions in the present section.

We have seen in Sect. 22 that, in order to determine in which Friedmann solution we live, it is necessary to determine the values, at the present epoch, of ρ (mass density of the galaxies) and α (expansion rate). It is quite clear how (at least in principle) one would determine the present value of ρ. One would select a spatial volume (much larger then the inter-galactic distance, so local irregularities would be averaged out), determine the total mass inside that volume, and divide the mass by the volume. This is the present value of ρ. But what measurements would one make on our Universe to determine the present value of α. We now begin the discussion of this – a much more complicated – question.

The following phenomenon is observed in our Universe. The light reaching us from a distant galaxy is found to have a longer wavelength than the light had when it left the galaxy. (We know the wavelength the light had when it left the galaxy because it arose from atomic transitions.) The red-shift of a galaxy is defined by $z = (\lambda_r - \lambda_e)/\lambda_e$, where λ_e is the wavelength of the light emitted by the galaxy, and λ_r is the wavelength of the light we receive. Thus, $z = 0$ corresponds to no shifting of wavelengths. It is found, furthermore, that the red-shift of a galaxy depends on the distance of that galaxy from us. Thus, experimentally, one determines a graph such as the one below.

This graph is called the *observed distance-red-shift relation.*

The idea is now the following. Fix a Friedmann solution. Then, using this solution, we can compute a predicted distance-red-shift relation. By comparing the various predicted relations with the actual observed relation, one can make statements

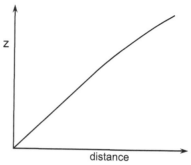

about which Friedmann solutions are active candidates to represent our own Universe. It will turn out that this comparison yields only one useful piece of information: the value of α at the present epoch. Thus, our next task is to compute, within a given Friemann solution, what the distance-red-shift relation should be.

Fix a world-line of dust to represent "our galaxy". Fix a point p on this world-line to represent "us, now". Draw a null geodesic from p into the past to some point q on the world-line of another dust particle. This null geodesic represents the light reaching us from the distant galaxy. The point q is the event "the galaxy emitted the light". Let k^a denote the tangent vector to this

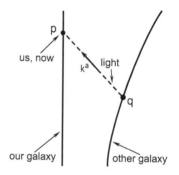

geodesic, so we have $k^a k_a = 0$ and $k^m \nabla_m k^a = 0$. This is the geometrical set-up. Two things must be discussed within this framework: the red-shifts, and the distances.

We begin with the red-shift. Let our light ray pass by a galaxy with four-velocity ξ^a. Then, as we saw in Sect.17, $\omega = k_a \xi^a$ is the (angular) frequency of the light as seen by an observer on the galaxy. In particular, ω, evaluated at q, is the emitted frequency of the light, while ω, evaluated at p, is the received frequency of the light. Let a prime denote $k^m \nabla_m$ (i.e., the derivative along the light ray). Then

$$\omega' = k^m \nabla_m \omega = -k^m \nabla_m (\xi^a k_a) = -\xi^a k^m \nabla_m k_a - k_a k^m \nabla_m \xi_a$$
$$= 0 - \alpha k^a k^b (g_{ab} + \xi_a \xi_b) = -\alpha k^a k_a - \alpha (k^a \xi_a)^2 = -\alpha \omega^2$$

where we have used (46). The meaning of this equation is clear physically. If α is positive, then the frequency of the light (as seen by intervening galaxies) decrease on moving from q to p. That is, the observed wavelength will be larger than the emitted wavelength.

Thus, if we have our Friedmann solution (so we know α everywhere), and if we know the frequency of the light emitted at q, then we can determine ω along our entire light ray from

$$\omega' = -\alpha\omega^2 \tag{65}$$

In particular, we can determine the value of ω at p, i.e., the frequency of the light received. These frequencies determine the corresponding wavelengths. Hence, from (65) we can determine the red-shift z for the other galaxy.

We now have half of what we need to draw the theoretical distance-red-shift relation. We know how to compute the redshift of the galaxy through event q. We next must determine the distance of that galaxy.

Now distance is not a "natural" concept in general relativity. That is, there is no simple, unambiguous quantity associated with the points p and q in the earlier figure which everyone would agree to call the "spatial distance" between the galaxies. This type of situation arises frequently in general relativity. What we must do is analyze in detail the actual measurements and computations the astronomer makes in arriving in the quantity he chooses to call the "distance" between the galaxies. We then express what the astronomer actually did in terms of tensors in our Friedmann solution. Thus, we acquire the ability to compute, for our given Friedmann solution, a quantity (associated with p and q) which can be called the distance between the galaxies. In practice, what the astronomer does to determine what he calls distance is rather complicated. It is convenient to over-simplify. We select one reasonable operational definition of distance, and analyze it within the Friedmann solutions.

In ordinary Euclidean space, if one looks at an object of actual size s, and if it subtends an angle θ (small), then we would say we 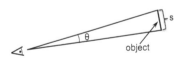 are a distance s/θ from the object. One might (and often does) use a similar technique in astronomy. One looks at a galaxy, knowing (say, from the study of galactic structure) its size s. One measures the apparent angular size of the galaxy, and thus computes its "distance". We shall adopt this as our definition of distance.

Thus, from the point "p" of our Friedmann solution, we draw two nearby null geodesics. These reach opposite sides of the galaxy located at q. Let s be the actual diameter of this galaxy. Our task is to compute θ, the angle we see these light rays reaching us (at p) with. Then s/θ will be the "distance" of this galaxy.

We are dealing with two nearby null geodesics. Clearly, therefore, the way to compute θ from s is to use the equation of geodesic deviation.

Let l^a be a vector, defined at each point of our null geodesic, which is parallel transported along the geodesic. That is, we have $k^m \nabla_m l_a = 0$ Hence,

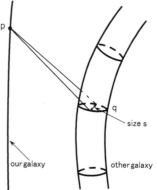

$$k^m \nabla_m (l^a l_a) = 2 l^a k^m \nabla_m l^a = 0$$
$$k^m \nabla_m (l^a h_a) = k_a (k^m \nabla_m l^a) + l^a (k^m \nabla_m k_a) = 0$$
$$k^m \nabla_m (l^a \xi_a) = \xi_a k^m \nabla_m l^a + l^a k^m \nabla_m \xi_a = 0 + l^a k^m \alpha (g_{am} + \xi_a \xi_m)$$
$$= \alpha (l^m k_m) + \alpha (\xi_a l^a)(\xi_m k^m)$$

Clearly, we can choose $l^a l_a = 1$ and $k^a l_a = 0$. We do so. We now ask that $\kappa \, l_a$ satisfy the equation of geodesic deviation, where κ is a function along the geodesic. That is, we ask that

$$k^m k^n \nabla_m \nabla_n (\kappa l_a) = -\kappa R_{manb} \, k^m k^n l^b \qquad (66)$$

Using $k^m \nabla_m l_a = 0$, the left side becomes $\kappa'' l_a$. To evaluate the right side, note that (50) together with isotropy implies

$$R_{abcd} = \frac{16}{7} \pi G \, \rho [g_{a[c} g_{d]b} - 3 \xi_{[a} g_{b][c} \xi_{d]}] \qquad (67)$$

Hence, the right side of (66) is $-4\pi G \rho \omega^2 \kappa \, l_a$. Getting rid of the l_a's, Eqn. (66) becomes

$$\kappa'' = -4\pi G \rho \omega^2 \kappa \qquad (68)$$

Geometrically, the situation is now the following. The vector l^a is just a unit direction. But κl^a satisfied the equation of geodesic deviation provided κ varies along our geodesic according to (68). Hence, κ represents the distance between the

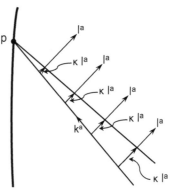

two nearby geodesics. We are concerned with the solution κ of (68) which satisfies the two boundary conditions: $\kappa(q) = s$ (the size of the other galaxy), and $\kappa = 0$ (the two nearby light rays meet at p). These two boundary conditions determine a unique solution of (68).

We now have a description of the two nearby light rays. Only one task remains; to determine the apparent angular size θ of the galaxy as seen from p. To do

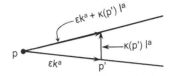

this, we first find the tangent vectors to the two null geodesics, at p. Move a parameter distance ϵ ($\ll 10$) along each geodesic from p. For the first geodesic, this infinitesimal displacement is represented by ϵk^a. For the second geodesic, the displacement is represented by $\epsilon k^a + \kappa(p') l^a$ where p' is the point displaced ϵk^a from p. (This is the definition of geodesic deviation.) But $\kappa(p) = 0$. Hence, $\kappa(p') = \epsilon k^m \nabla_m \kappa$ (at p) $= \epsilon \kappa'(p)$. Hence, the tangent vectors to the two light rays are k^a and $k^a + \kappa'(p) l_a$. To obtain the apparent spatial direction from which we (at p) see these light rays coming in, we project orthogonal to ξ^a (the four-velocity of our galaxy). Hence, the light rays appear to come in in spatial directions $h_a{}^m k_m$ and $h_a{}^m(k_m + \kappa'(p) l_m) = h_a{}^m k_m + \kappa'(p) l_a$. What is the angle between these two nearby spatial vectors? From Euclidean geometry, it is

$$\theta = [h^{am} k_m h_a{}^n k_n]^{-1/2} [\kappa'(p) l_a \kappa'(p) l^a]^{1/2} = \frac{\kappa'(p)}{\omega(p)} \tag{69}$$

We are now prepared to summarize what is a rather complicated situation. We first pick a number s to represent the size of the galaxy at q. We then solve (68) for κ along our null geodesic,

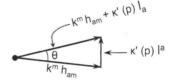

subject to $\kappa(p) = 0$ and $\kappa(q) = s$. Substituting κ' into (69), we obtain the angle subtended by the galaxy. Then s/θ is what we call the distance of that galaxy. (Note that this distance is independent of s. Since (68) is linear, the function κ defined by (68) is proportional to s. Hence, from (69), θ is proportional to s. So, s/θ is independent of s. This, of course, is what we would expect.)

We now have a prescription for computing the theoretical distance-red-shift relation. Choose a point p (to represent us, now), and a null geodesic into the past from p to some point q. Solve (65) to determine the redshift of the light from this galaxy. Then pick s (it makes no difference what s), and solve (68) subject to $\kappa(p) = 0$ and $\kappa(q) = s$. Find θ from (69), and set distance $= s/\theta$. Thus, once we pick q, we can compute the redshift z and the distance s/θ. Repeating for various choices of q (the event of light leaving the other galaxy), we obtain the theoretical distance-red-shift relation.

In fact, we shall not attempt to solve these equations. (The solutions are complicated, not very illuminating, and not very useful in practice.) Instead, we ask a simpler and more important question. What is the slope of the distance-red-shift curve at the origin (i.e., the value of $dz/d(\text{distance})$ at $z = 0$? The observed distance-red-shift curve is nearly a straight line, so the only quantity which has been determined observationally with any accuracy is this slope. Let us compute the slope of the theoretical distance-red-shift curve.

Let q lie a parameter-amount ϵ ($\epsilon \ll 1$) along the null geodesic from p, so the connecting vector is ϵk^a. Then since $\kappa(p) = 0$, $\kappa(q) = \epsilon \kappa'(p)$, which is that we call s. Hence, from (69), $\theta = \kappa'(p)/\omega(p) = (s/\epsilon)/\omega(p)$. So, distance $s/\theta = \epsilon \omega(p)$. (Note that, to first order, we don't even need to use (68).) Thus, the distance between the two galaxies in this case is $\epsilon\omega(p)$. We now want to compute the redshift between these two galaxies. From (65), $\omega(p) = \omega(q) - \epsilon \alpha(p)\,\omega^2(p)$, to first order in ϵ. Hence,

$$z = \frac{\lambda_{rec} - \lambda_{emit}}{\lambda_{emit}} = \frac{1/\omega(p) - 1/\omega(q)}{1/\omega(q)} = \frac{\omega(q)}{\omega(p)} - 1$$

$$= \frac{\omega(p) + \epsilon\,\alpha(p)\,\omega^2(p)}{\omega(p)} - 1 = \epsilon\,\alpha(p)\,\omega(p)$$

to first order in ϵ. Taking the ratio, $z/\text{distance} = \alpha(p)$. Thus, *the slope of the theoretical distance-red-shift curve at the origin is precisely the value of α at the observation point p.* (It's clear, from the definition of α and the remark surrounding it, that this is the only reasonable answer. This is why we called α the "rate of expansion".)

Thus, the value of α in our Universe at the present epoch is determined merely by measuring the slope of the distance-red-shift curve (observational) at the origin.

We can now determine the two quantities – ρ and α at the present epoch – necessary to determine the Friedmann solution in which we live. What do the observations say? Firstly, this α – called the *Hubble constant* – is known relatively well. The best current value is $\alpha = 1.7 \times 10^{-19}\,\text{sec}^{-1}$ (i.e., the Universe expands one part in 2×10^{11} per year). That's good perhaps to a factor of five. The current value of ρ is known much less accurately. What is the most interesting question? Clearly, it's whether the constant c in (63) is positive, negative, or zero (i.e., whether our Universe is closed spatially and will later contract, or whether it is open and will continue expanding). Set

$$\rho_c = \frac{\alpha^2}{\frac{8}{3}\pi G} = 170 \text{ g/sec}^3 \quad (= 6.3 \times 10^{-30} \text{ g/cm}^3) \tag{70}$$

the critical density. Then, if ρ (observed) $> \rho_c$, we have the spatially closed Universe, while if ρ (observed) $< \rho_c$, we have the open Universe.

The best current value for ρ is $\rho = (1/30)\rho_c$. This value suggests the spatially open, ever expanding Universe. Unfortunately, the observational error in this ρ is such that no definite conclusion can be drawn, at the present time, on this question.

It's all very unsatisfactory. General relativity makes very definite, detailed predictions. Any reasonable theory would give predictions of the same general order of magnitude. The observations, as a whole, are good to factors such as ten. We cannot even make a definite statement, from the observations, as to whether our Universe should be spatially open or closed.

24. Symmetries

The notion of a symmetry – a motion in space-time under which all the physics is invariant – is an important one. It is often the case in practice that essentially all the simple and useful information about a space-time arises from the presence of symmetries.

Recall the motion of the Lie derivative. We have a vector field σ^a on a manifold M. Suppose we displace each point of M infinitesimally by an amount determined by the value of σ^a at

that point. Thus, σ^a defines a "motion" – an infinitesimal "sliding along" in the direction of σ^a. Now let $T^{a\cdots c}{}_{b\ldots d}$ be any tensor field in M. Then one obtain two tensors fields: the original $T^{a\cdots c}{}_{b\ldots d}$ and the field which results from subjecting $T^{a\cdots c}{}_{b\ldots d}$ to this "sliding along" by σ^a. The second field minus the first, divided by the (parameter) amount of sliding along, gives the *Lie derivative* of $T^{a\cdots c}{}_{b\ldots d}$. It follows that the Lie derivative of $T^{a\cdots c}{}_{b\ldots d}$ by σ^a is given by the right side of

$$\mathscr{L}_\sigma T^{a\cdots c}{}_{b\ldots d} = \sigma^m \nabla_m T^{a\cdots c}{}_{b\ldots d} - T^{m\cdots c}{}_{b\ldots d} \nabla_m \sigma^a - \cdots$$
$$- T^{a\cdots m}{}_{b\ldots d} \nabla_m \sigma^c + T^{a\cdots c}{}_{m\ldots d} \nabla_b \sigma^m + \cdots + T^{a\cdots c}{}_{b\ldots m} \nabla_d \sigma^m \quad (71)$$

A vector field σ^a in space-time is said to define a *symmetry* if all the tensor fields of physical interest on the space-time (i.e., the metric, and the fields which describe the various types of matter in the space-time) are invariant under the "sliding" by σ^a. That is, σ^a defines a symmetry if $\mathscr{L}_\sigma$ (all physical fields)$= 0$.

In particular, if σ^a is a symmetry, then $\mathscr{L}_\sigma g_{ab} = \sigma^m \nabla_m g_{ab} + g_{mb} + g_{am}\sigma^m = 2\nabla_{(a}\sigma_{b)}$. This equation,

$$\nabla_{(a}\sigma_{b)} = 0 \quad (72)$$

is called *Killing's equation*, the solutions *Killing vectors*. (Thus, a symmetry vector field is always a Killing vector (field). It usually turns out that Killing vectors are also symmetries.)

Note, that, since (72) is linear, any linear combination of Killing vectors (with constant coefficients) is again a Killing vector. Another elementary consequence of Killing's equation is the following . Let σ^a be a Killing vector, and set $\kappa = \sigma^b \sigma_b$. Then

$$\sigma^a \nabla_a \kappa = -\sigma^a \nabla_a (\sigma^b \sigma_b) = 2\sigma^a \sigma^b \nabla_a \sigma_b$$
$$= 2\sigma^a \sigma^b \nabla_{(a} \sigma_{b)}$$

where we have used (72) in the last step. That is, the norm of a Killing vector is constant along the integral curves of σ^a. Furthermore, if σ^a is a Killing vector, then $\sigma^m \nabla_m \sigma_a = -\sigma^m \nabla_a \sigma_m = -\frac{1}{2} \nabla_a \kappa$, where we have used (72) in the first step. Thus, the curvature of an integral of a Killing vector is $-\frac{1}{2}$ times the gradient of the norm κ. (Therefore: Lemma: The integral curves of a Killing vector are geodesics when and only when the Killing vector has constant norm.)

We now derive what is perhaps the most important single consequence of (72). Let σ^a be a Killing vector. Then

$$\nabla_a \nabla_b \sigma_c = \nabla_a \nabla_c \sigma_b = -\nabla_c \nabla_a \sigma_b - R_{acb}{}^m \sigma_m$$
$$= \nabla_c \nabla_b \sigma_a - R_{acb}{}^m \sigma_m = \nabla_b \nabla_c \sigma_a + R_{cba}{}^m \sigma_m - R_{acb}{}^m \sigma_m$$
$$= -\nabla_b \nabla_a \sigma_c + R_{cba}{}^m \sigma_m - R_{acb}{}^m \sigma_m$$
$$= -\nabla_a \nabla_b \sigma_c - R_{bac}{}^m \sigma_m + R_{cba}{}^m \sigma_m - R_{acb}{}^m \sigma_m$$

where, in the first, third and fifth steps, we have used (72), and in the second, fourth, and sixth steps we have used the definition of the Riemann tensor. Adding $\nabla_a \nabla_b \sigma_c$ to both sides (i.e., to the first and last expression) of this equation, and using $R_{acb}{}^m + R_{cba}{}^m + R_{bac}{}^m = 0$ (a Riemann tensor symmetry) we obtain $\nabla_a \nabla_b \sigma_c = 2R_{cba}{}^m \sigma_m$. But $R_{cba}{}^m = R^m{}_{abc}$ (Riemann tensor symmetry). Hence, we obtain

$$\nabla_a \nabla_b \sigma_c = R^m{}_{abc} \sigma_m \tag{73}$$

This is the desired result. (Note that the result of antisymmetrizing (73) over "a" and "b" is an identity.) That is, if the first derivative of a vector field is restricted by (72), then its second derivative can be expressed (via (73)) in terms of the original vector field. We need never deal directly with second or higher derivatives of a Killing vector.

We now derive two consequences of (73), Let σ^a and γ^a be Killing vectors, and set

$$\rho^a = [\sigma, \gamma]^a = \sigma^m \nabla_m \gamma^a - \gamma^m \nabla_m \sigma^a \tag{74}$$

We show that ρ^a is also a Killing vector. Taking a derivative of ρ^b:

$$\nabla^a\rho^b = (\nabla^a\sigma^m)(\nabla_m\gamma^b) + \sigma^m\nabla^a\nabla_m\gamma^b - (\nabla^a\gamma^m)(\nabla_m\sigma^b) - \gamma^m\nabla^a\nabla_m\sigma^b$$
$$= [(\nabla^a\sigma^m)(\nabla_m\gamma^b) - (\nabla^a\gamma^m)(\nabla_m\sigma^b)] + [\sigma^m\nabla^a\nabla_m\gamma^b - \gamma^m\nabla^a\nabla_m\sigma^b]$$
$$= [-(\nabla^m\sigma^a)(\nabla_m\gamma^b) + (\nabla^m\gamma^a)(\nabla_m\sigma^b)] + [\sigma^m R^{na}{}_m{}^b\gamma_n - \gamma^m R^{na}{}_m{}^b\sigma_n]$$
$$= [-(\nabla^m\sigma^a)(\nabla_m\gamma^b) + (\nabla^m\sigma^b)(\nabla_m\gamma^a)] + [\sigma^m\gamma^n R_n{}^a{}_m{}^b - \sigma^m\gamma^n R_m{}^b{}_n{}^a]$$

where, in the first step, we have substituted (74); in the second, we have rearranged terms; in the third, we have used (72) in the first bracket and (73) in the second; and in the fourth we have reversed the roles of (the dummy indices) "m" and "n" and used $R_{abcd} = R_{cdab}$ in the last term in the second bracket. Now consider the right side of this expression. Each bracket is, evidently, antisymmetric in "a" and "b". Hence, $\nabla^{(a}\rho^{b)} = 0$, so ρ^a is a Killing vector.

The vector field $[\sigma, \gamma]^z$ is called the *Lie Bracket*, or *commutator*, of σ^a and γ^a. Evidently, $[\sigma, \gamma]^a = -[\gamma, \sigma]^a$. Furthermore , the Lie bracket is linear (using linear combinations with constant coefficients) in σ^a and γ^a. Finally, it is not difficult to check that

$$[[\sigma, \gamma], \lambda]^a + [[\gamma, \lambda], \sigma]^a + [[\lambda, \sigma], \gamma]^a = 0$$

(The remarks to here in this paragraph are true for any vector fields – whether or not they are Killing vectors.) Thus, the Killing vectors on a space-time have the structure of a Lie algebra (a vector space with an antisymmetric bracket operation, subject to the Jacobi identity).

As a second example of the use of (73), we find all Killing vectors in flat space (of any signature and dimension). Let σ^a be a Killing vector. Then, since $R_{abcd} = 0$ (flat space), (73) gives $\nabla_b\sigma_c = 0$. Hence,

$$\nabla_a\sigma_b = F_{ab} \tag{75}$$

where F_{ab} is a constant tensor field. Eqn. (72) implies that F_{ab} is antisymmetric. To integrate (75), introduce an origin O in our flat space, and let x^a be the vector field whose value at each point is the position vector of that point relative to O. (Precisely, x^a is defined by the properties that x^a vanishes at O, and $\nabla_a x^b = \delta_a{}^b$.) The general solution of (75) is thus

$$\sigma_b = x^a F_{ab} + \underline{\sigma}_b \tag{76}$$

where $\underline{\sigma}_b$ is a constant vector field. Thus, the general Killing vector in flat space is defined by a constant antisymmetric tensor field F_{ab} and a constant vector field $\underline{\sigma}_b$. (The particular constant fields which represent a given Killing vector depend on the choice of origin.)

The Killing vectors $\sigma^a = \underline{\sigma}^a$ are called translations. If our flat space is Minkowski space, a translation is said to be temporal or spatial according as $\underline{\sigma}^a$ is timelike or spacelike. (Of course, $\underline{\sigma}^a$ could also be null in Minkowski space.) Consider Eucliden 3-space. Then $\underline{F}_{ab} = \epsilon_{abc}W^c$, for some constant vector field W^c. The corresponding Killing vector, $\sigma_b = \underline{F}_{ab}x^a$, is called a rotation about the origin O, with axis of rotation W^a. Consider Minkowski space. Then $\sigma_b = \underline{F}_{ab}x^a$ is said to generate a Lorentz transformation about O. To decompose Lorentz transformations, introduce a unit timelike vector t^a at O. Then, since $\underline{F}_{ab}$ is antisymmetric, we can introduce the "electric" and "magnetic" parts of $\underline{F}_{ab}$, with respect to t^a. (The equations are precisely (20), (21), and (22)). The electric part of $\underline{F}_{ab}$, $t^b\underline{F}_{ab}$ is said to generate a boost, or velocity transformation, in the direction $t^b\underline{F}_{ab}$. The magnetic part is said to generate a spatial rotation about the axis $\frac{1}{2}\epsilon_{abcd}t^b\underline{F}^{cd}$. Thus, the decomposition of Lorentz transformations into boosts and rotations depends on the choice of a time-direction t^a.

A vector field σ_a (not necessarily a Killing vector) in space-time is said to be *hypersurface-orthogonal* if

$$\sigma_{[a}\nabla_b\sigma_{c]} = 0 \tag{77}$$

Geometrically, (77) is the condition that there exist a family of three-dimensional surfaces in space-time to which σ_a is everywhere orthogonal. (The fibers of a twisted rope define a vector fields in Euclidean 3-space which is not hypersurface-orthogonal. Translations are always hypersurface-orthogonal.)

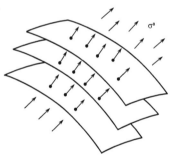

We let σ^a be a hypersurface-orthogonal Killing vector, i.e., let σ^a satisfy (72) and (77). (About half of all Killing vectors one meets in practice are hypersurface-orthogonal.) Contracting (77) with σ^a, using $\sigma^a\nabla_a\sigma_b = -\sigma^b\nabla_b\sigma_a = \frac{1}{2}\nabla_a\kappa$, where, $\kappa = \sigma^s\sigma_c$, we obtain (when $\kappa \neq 0$)

$$\nabla_a\sigma_b = -\kappa^{-1}\sigma_{[a}\nabla_{b]}\kappa \tag{78}$$

Thus, the value of the derivative of a hypersurface-orthogonal Killing vector at a point is expressible in terms of the values of the Killing vector, its norm, and the gradient of its norm at the point. This equations, and the result of substituting it into (73), are frequently the keys to unraveling a space-time with a hypersurface-orthogonal Killing vector.

All we have done so far in this section is introduce some definitions and some useful mathematical techniques. What is the physics of Killing vectors? We have seen that Killing vectors are associated with symmetries, and of course symmetries are usually associated with conserved quantities. So Killing vectors should lead to conserved quantities. These come in two varieties.

Let σ^a be a Killing vector in space-time, and consider a geodesic with tangent vector p^a. Then

$$p^m \nabla_m (p^a \sigma_a) = \sigma_a p^m \nabla_m p^a + p^a p^m \nabla_m \sigma_a$$
$$= 0 + p^a p^m \nabla_{(m} \sigma_{s)} = 0$$

Hence, $p_a \sigma^a$ is constant along geodesic. That is, $p_a \sigma^a$ is a "constant of the motion of the geodesic". (Not a very good term, since geodesics do not "move". (Nothing "moves" in space-time.)) (It is easily checked that if $p_a \sigma^a$ is constant along the geodesic, for every geodesic, then σ^a is a Killing vector.)

The second type of conserved quantity arises from the stress-energy T^{ab}. Let σ^a be a Killing vector. Then

$$\nabla_a (T^{ab} \sigma_b) = \sigma_b \nabla |a T^{ab} + T^{ab} \nabla_a \sigma_b = 0$$

That is, $T^{ab} \sigma_b$ is divergence-free. Now suppose we have a star, or some isolated body. We integrate $T^{ab} \sigma_b$ over a three-dimensional surface S cutting the world-tube of the body. If we were to integrate instead over some other surface S', then the difference between the integrals would be

$$\int_{S'} T^{ab} \, dS_a - \int_S T^{ab} \sigma_b \, dS_a = \int_V \nabla_a (T^{ab} \sigma_b) \, dV = 0$$

where V is the portion of the world-tube between S and S', Hence, $\int_S T^{ab} \sigma_a \, dS_a$ is just a number, independent of the surface S. It is a conserved quantity ("conserved" in the sense that the value of the integral is independent of the surface).

In a general space-time, there will be no Killing vectors (as we would expect: most systems have no symmetries). Killing vectors are important, not because they arise commonly in actual systems, but because permit us to

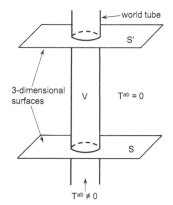

do something (e.g., with Einsten's equation) where otherwise we might be able to do nothing. The neat decomposition of Killing vectors in Minkowski space into translations, rotations, and boosts is not available for Killing vectors in a curved space-time. A Killing vector is just a Killing vector. However, often a Killing vector in a curved space-time closely resembles, in certain respects, a Killing vector in Minkowski space. (This is usually the case when the space-time is asymptotically flat, so one can examine the asymptotic behavior of the Killing vector.) When such a resemblances exists, it is usually emphasized by calling the Killing vector by the name of its Minkowski space analog (e.g., calling it a temporal translation, a rotation, etc.).

If a Killing vector in space-time is called a temporal translation, then the conserved quantity $p_a \sigma^a$ (where p_a is the four-momentum of a particle on the geodesic) is called the energy of the particle, while the conserved quantity associated with T^{ab} is called the energy of the star. Similarly, the conserved quantities associated with Killing vectors called spatial translations are called components of the momentum, and the conserved quantities associated with Killing vectors called spatial rotations are called components of the angular momentum. (For some reason, the conserved quantities associated with boosts do not seem to be very useful. They have no name.) Examples: Consider Minkowski space, with all its Killing vectors. Then what we above called the energy, momentum, and angular momentum associated with a p^a or T^{ab} are precisely what is called energy, momentum, and angular momentum in special relativity. (Note that these concepts from special relativity are, in a sense, simpler when introduced from the point of view of general relativity. It is clear why conserved quantities arise (because of symmetry). Furthermore, we don't have to prove that things "transform like a 4-vectors".)

Certain intuitive "symmetry expressions" are given a precise meaning in general relativity. We introduce four such. A space-time is said to be "stationary" if it possesses a Killing vector which is everywhere timelike. A space-time is said to be "static" if it possesses a Killing vector which is hypersurface-orthogonal and everywhere timelike. (Physically, a stationary system presents the same aspect at each time, although there may be motion (e.g., water flowing through a pipe, or a rotating disc), while in a static system nothing moves.) A space-time is said to be "axially symmetric" if it possesses a Killing vector which is everywhere spacelike, and whose integral curves are closed curves. (The axis is the points of space-time where this Killing vector vanishes. It is normally a two-surface – but it is possible that there exist no such points!) Finally, a space-time is said to be "spherically symmetric" if it possess three Killing vectors, $\underset{1}{l}{}^{a}$, $\underset{2}{l}{}^{a}$, and $\underset{3}{l}{}^{a}$ (physically, which generate rotations about the three axes), all of which are everywhere spacelike, which are linearly dependent at each point (i.e., $l^{[a}_1 l^b_2 l^{c]}_3 = 0$), and which have the same commutation relations as do the generators of the rotation group:

$$[l_1, l_2]^a = l_3^a, \quad [l_2, l_3]^a = l_1^a, \quad [l_3, l_1]^a = l_2^a \tag{79}$$

(It is easily checked that the three Killing vectors which generate rotations about an origin in Euclidean three-space have these properties.)

There is a patters to these definitions. In each case, a term means the existence of Killing vectors which display as many features reminiscent of that term as are simple and meaningful in curved space-time.

25. The Schwarzschild Solution

Physically, the Schwarzschild solution represents the geometry of an "isolated, non rotating star, which has settled down to equilibrium". What properties would we expect such a solution to have? Firstly, we would expect the solution to be static, i.e., we would expect to have a timelike, hypersurface-orthogonal Killing vector t^a. Secondly, we would expect the solution to be spherically symmetric, i.e., we would expect to have Killing vectors $\underset{1}{l^a}$, $\underset{2}{l^a}$, and $\underset{3}{l^a}$ which are spacelike, linearly dependent, and have the commutation relations (79). Finally, we would expect that the time-translations and rotations commute, i.e., we would expect to have additional commutation relations

$$[t, \underset{1}{l}]^a = [t, \underset{2}{l}^a] = [t, \underset{3}{l}^a] = 0 \qquad (80)$$

To summarize, we are concerned with space-time having four Killing vectors, with the commutation relations (79) and (80). For the matter composing the star, we take a fluid. Thus, we have the density ρ, pressure p, and (unit) velocity field η^a. Since the star is supposed to have "settled down to equilibrium", we suppose that the fluid does not "move relative to static observers", i,e., we take η^a a multiple of t^a.

To summarize, the Schwarzschild solution is a space-time with four Killing vectors, t^a (timelike, hypersurface-orthogonal), and $\underset{1}{l^a}$, $\underset{2}{l^a}$, $\underset{3}{l^a}$ (spacelike, linear dependent), subject to (79) and (80), where the matter is a fluid with four-velocity field proportional to t^a. We now discuss the geometry of the Schwarzschild solution.

Set $\lambda = t^a t_a$, (so $\lambda < 0$ since t^a is timelike). We begin with the physical interpretation of this λ. Since t^a is hypersurface-orthogonal, we have, from (78)

$$\nabla_b t_c = -\lambda^{-1} t_{[b} \nabla_{c]} \lambda \qquad (81)$$

Hence,

$$R^m{}_{abc}\, t_m = \nabla_a \nabla_b t_c = \nabla_a(-\lambda^{-1}\, t_{[b}\nabla_{c]}\lambda)$$
$$= \lambda^{-2}(\nabla_a\lambda)\, t_{[b}\nabla_{c]}\lambda - \lambda^{-1}(\nabla_a\, t_{[b})\nabla_{c]}\lambda - \lambda^{-1}(\nabla_a\nabla_{[c}\lambda)t_{b]}$$
$$= \frac{1}{2}\lambda^{-2}\nabla_a\lambda\, t_{[b}\nabla_{c]}\lambda - \lambda^{-1}(\nabla_a\nabla_{[c}\lambda)t_{b]}$$

where, in the first step, we have used (73); in the second, we have substituted (81); in the third, we have expanded using the Leibniz rule, and in the fourth, we have again used (81).

Contracting the previous equation over "a" and "c",

$$R^m{}_b\, t_m = \frac{1}{2}\lambda^{-2}(\nabla^c\lambda)t_{[b}\nabla_{c]}\lambda - \lambda^{-1}(\nabla^c\nabla_{[c}\lambda)t_{b]}$$
$$= \frac{1}{4}\lambda^{-2}(\nabla^c\lambda)(t_b\nabla_c\lambda - t_c\nabla_b\lambda) - \frac{1}{2}\lambda^{-1}((\nabla^c\nabla_c\lambda)t_b - (\nabla^c\nabla_b\lambda)t_c)$$
$$\tag{82}$$
$$= \frac{1}{4}\lambda^{-2}t_b(\nabla^c\lambda\nabla_c\lambda) - \frac{1}{2}\lambda^{-1}t_b\nabla^2\lambda + \frac{1}{2}\lambda^{-1}t^c\nabla_c\nabla_b\lambda$$

where, in the second step, we have expanded the antisymmetrizations, and in the third step, we have used $t^c\nabla_c\lambda = 0$ and written ∇^2 for $\nabla^c\nabla_c$. The last term on the right in (82) can be simplified further:

$$t^c\nabla_c\nabla_b\lambda = t^c\nabla_b\nabla_c\lambda = \nabla_b(t^c\nabla_c\lambda) - (\nabla_c\lambda)\nabla_b t^c$$
$$= \nabla_b(0) - \nabla^c\lambda\nabla_b t_c = -\nabla^c\lambda(-\lambda^{-1}t_{[b}\nabla_{c]}\lambda)$$
$$= \frac{1}{2}\lambda^{-1}t_b(\nabla^c\lambda\nabla_c\lambda)$$

Hence, (82) becomes

$$R_{mb}\, t^m = \frac{1}{2}\lambda^{-2}t_b(\nabla^c\lambda\nabla_c\lambda) - \frac{1}{2}\lambda^{-1}t_b\nabla^2\lambda \tag{83}$$

Eqn. (83) gives components of the Ricci tensor in terms of λ. We proceed (as always in such situations) to eliminate the Ricci tensor in favor of the matter variables. For a perfect fluid, Einstein's equation becomes

$$R_{ab} = 8\pi\, G[-\lambda^{-1}(\rho + p)t_a t_b + \frac{1}{2}(\rho - p)g_{ab}] \tag{84}$$

where we have used the fact that the velocity field of the fluid is $(-\lambda)^{-1/2}t^a$ (i.e., that multiple of t^a which is unit). From (84), $R_{ab}\, t^b = -4\pi\, G\, t_a(\rho + 3p)$. Substituting, we have from (83)

$$\lambda^{-1}\nabla^2\lambda - \lambda^{-2}(\nabla^2\lambda\nabla_c\lambda) = 8\pi\, G(\rho + 3p) \tag{85}$$

This can be rewritten in the more suggestive form

$$\nabla^2[\frac{1}{2}\ln(-\lambda)] = 4\pi\, G(\rho + 3p) \tag{86}$$

Eqn. (86) has a simple physical interpretation. We regard $\frac{1}{2}\ln(-\lambda)$ as analogous to the "Newtonian potential " φ, so (86) in analogous to the Newtonian equation $\nabla^2\varphi = 4\pi\, G\rho$. Note that it is $(\rho + 3p)$, rather than just ρ, which appears on the right in (86). We may regard $(\rho + 3p)$ as representing the "active gravitational mass density" (i.e., the quantity which describes the ability to make gravitational field) of the fluid.

Further support for this interpretation comes from a simple calculation. Set $\eta^a = (-\lambda)^{-1/2}t^a$ (so η^a is the unit four-velocity of an observer following an integral curve of t^a, the time-translation). Then

$$\begin{aligned}
\eta^b\nabla_b\eta^a &= \eta^b\nabla_b[(-\lambda)^{-1/2}t^a] = t^a\eta^b\nabla_b(-\lambda)^{-1/2} + (-\lambda)^{-1/2}\eta^b\nabla_b t^a \\
&= 0 + (-\lambda)^{-1/2}\eta^b\nabla_b t^a = -\lambda^{-1}t^b\nabla_b t^a = \lambda^{-1}t^b\nabla^a t_b \tag{87} \\
&= \frac{1}{2}\lambda^{-1}\nabla^a(t^b t_b) = \frac{1}{2}\lambda^{-1}\nabla^a\lambda = \nabla^a[\frac{1}{2}\ln(-\lambda)]
\end{aligned}$$

Thus, the acceleration felt by this observer is $\nabla^a[\frac{1}{2}\ln(-\lambda)]$. Eqn. (87) is analogous to the Newtonian equation $\vec{A} = -\vec{\nabla}\varphi$ for the acceleration felt by a free particle in potential φ. Why the difference in sign between the Newtonian expression and (87)? Because of the difference in the meaning of "acceleration" in the two cases. In the Newtonian case, we release the particle, and observe its acceleration (relative to, e.g., the star) toward the star. In general relativity, such a particle would be said to have zero acceleration (since acceleration is measured by the cube and springs). Instead we hold the particle at fixed distance from the star, and then note its acceleration (measured by the cube). In the Newtonian case, the free particle "accelerates" toward the star; in general relativity, the held particle is (since it is held) accelerated away from the star.

The remarks above are intended to motivate and to give a feeling for the calculations. We now discuss the geometry in detail.

We first obtain some additional facts about the Killing vectors. We have

$$\underset{1}{l}^a\nabla_a\lambda = \underset{1}{l}^a\nabla_a(t^b t_b) = 2t^b\underset{1}{l}^a\nabla_a t^b = 2t^b t^a\nabla_a\underset{1}{l}_b = 0$$

where, in the second step, we have expanded using the Leibniz rule, in the third step, we have used $[l^a, t]^a_1 = 0$, and, in the fourth step, we have used $\nabla_{[a} l_{b]}_1 = 0$. In words, λ *is invariant under the rotations.* (This, of course, is what we would expect.) We next prove that *the rotational Killing vectors are orthogonal to the timelike Killing vector.* The proof is straightforward:

$$
\begin{aligned}
t^a l_a &= t^a (l^b \nabla_b l_a - l^b \nabla_b l_a) = -t^a l^b \nabla_a l_b + t^a l^b \nabla_a l_b \\
&= -l^a l^b \nabla_a t_b + l^a l^b \nabla_a t_b = 2l^a l^b \nabla_a t_b \\
&= 2l^a l^b (-\lambda^{-1} t_{[a} \nabla_{b]} \lambda) = 0
\end{aligned}
$$

where, in the first step we have used $l_a = [l, l]_a$; in the second, Killing's equation on l_a and l_a; in the third, $[l, t]^a = [l, t]^a = 0$; in the fourth, Killing's equation on t_a; in the fifth, (81); and, in the sixth step, the fact that $l^a \nabla_a \lambda = l^a \nabla_a \lambda = 0$. It is perhaps not surprising that "rotation" gives a motion in a space-time orthogonal to the motion "time-translation".

We introduce a (positive) scalar field r in space-time:

$$
2r^2 = l^a l_a + l^a l_a + l^a l_a \tag{88}
$$

We interpret r as representing a sort of "radial distance from the center of the star", an interpretation justified by the observation that this is precisely what the r given by (83) is for rotational Killing vectors in Euclidean space. We next show that r *is invariant under the time-translation and the rotations* (more precisely, $t^a \nabla_a r = l^a \nabla_a r = l^a \nabla_a r = l^a \nabla_a r = 0$) (Note, again, that this assertion is obvious geometrically.) First, the time-translation part:

$$
t^a \nabla_a (l^b l_b) = 2t^a l^b \nabla_a l_b = 2l^a l^b \nabla_a t_b = 0
$$

where, in the second step, we have used $[t, l]^a = 0$. repeating for $l^b l_b$ and $l^b l_b$, and summing, using (88), we obtain $t^a \nabla_a r = 0$. For the rotation part:

$$\underset{1}{l^a}\nabla_a(2r^2) = \underset{1}{l^a}\nabla_a(\underset{1}{l^b}\underset{1}{l_b} + \underset{2}{l^b}\underset{2}{l_b} + \underset{3}{l^b}\underset{3}{l_b})$$

$$= 2\underset{1}{l^a}\underset{1}{l^b}\nabla_a\underset{1}{l_b} + 2\underset{1}{l^a}\underset{2}{l^b}\nabla_a\underset{3}{l_b} + 2\underset{1}{l^a}\underset{3}{l^b}\nabla_a\underset{2}{l_b}$$

$$= 0 + 2\underset{2}{l^b}(\underset{2}{l^a}\nabla_a\underset{1}{l_b} + \underset{3}{l_b}) + 2\underset{3}{l^b}(\underset{3}{l^a}\nabla_a\underset{1}{l_b} - \underset{2}{l_b})$$

$$= 2\underset{2}{l^b}\underset{3}{l_b} - 2\underset{3}{l^b}\underset{2}{l_b} = 0$$

where, in the third step, we have used (79), and, in the fourth step, we have used Killing's equation on $\underset{1}{l_b}$.

Finally, we introduce the scalar field $\mu = (\nabla^a r)\nabla_a r$. In flat space, $\mu = 1$ (since r is distance). Hence, deviations of μ from this value represent "curvature of space". By an argument similar to those above (or, using properties of Lie derivatives, in one step) we have that μ *is invariant under the time-translation and rotations.*

Let us summarize the situation. We think of r as a "radial coordinate". We think of λ and μ as "fields which describe the geometry of space-time". Since our space-time is static and spherically symmetric, we expect that everything of interest will be a function only of r. This expectation is made precise as follows. We have seen that the derivative of any of r, μ or λ in the direction of any of t^a, $\underset{1}{l^a}$, $\underset{2}{l^a}$, or $\underset{3}{l^a}$ is zero. Since t^a, $\underset{1}{l^a}$, $\underset{2}{l^a}$, and $\underset{3}{l^a}$ are linearly dependent, and t^a is orthogonal to all three, these four vectors span a three-dimensional vector space at each point of space-time. Hence, there is only one direction, at each point, simultaneously orthogonal to t^a, $\underset{1}{l^a}$, $\underset{2}{l^a}$, and $\underset{3}{l^a}$. But $\nabla_a r$, $\nabla_a \lambda$ and $\nabla_a \mu$ are simultaneously orthogonal to these vectors. Hence, $\nabla_a r$, $\nabla_a \lambda$ and $\nabla_a \mu$ are all proportional to each other. In other words, $\lambda = \lambda(r)$ and $\mu = \mu(r)$, functions of r.

The idea is to use Einstein's equation to obtain a pair of ordinary differential equations on the functions $\lambda(r)$ and $\mu(r)$ We have one such equation, (85). Let us work in the region outside the star ($\rho = p = 0$). Then (85) becomes

$$\nabla^2 \lambda - \lambda^{-1}(\nabla^a \lambda)(\nabla_a \lambda) = 0 \tag{89}$$

We need a second equation. From the definition of the Riemann tensor, $\nabla_{[a}\nabla_{b]}\nabla_c r = \frac{1}{2}R_{abc}{}^m \nabla_m r$. Contracting over "a" and "c" and then contracting with $\nabla^b r$ we obtain

$$\frac{1}{2}\nabla^2(\nabla^a r \nabla_a r) - (\nabla^a \nabla^b r)(\nabla_a \nabla_b r) - \nabla^a r \nabla_a(\nabla^2 r)$$

$$= R_{ab}\nabla^a r \nabla^b r = 4\pi\, G\, \mu(\rho - p)$$

where we have eliminated the Ricci tensor using (84). Hence,

$$\frac{1}{2}\nabla^2(\nabla^a r \nabla_a r) - (\nabla^a \nabla^b r)(\nabla_a \nabla_b r) - \nabla^a r \nabla_a \nabla^2 r \qquad (90)$$

Here is our second piece of Einstein's equation. All that remains is to rewrite (89) and (90) as ordinary differential equations on $\lambda(r)$ and $\mu(r)$.

Denote d/dr by a prime. Then $\nabla_a \lambda = \lambda' \nabla_a r$ (chain rule). Hence, $\nabla^2 \lambda = (\nabla^a \lambda') \nabla_a r + \lambda' \nabla^2 r = \lambda'' \nabla^a r \nabla_a r + \lambda' \nabla^2 r = \lambda'' \mu + \lambda' \nabla^2 r$. Similarly, $\nabla^2 \mu = \mu'' \mu + \mu' \nabla^2 r$. Hence, (89) and (90) become

$$\lambda'' \mu + \lambda' \nabla^2 r - \lambda^{-1}(\lambda')^2 \mu = 0 \qquad (91)$$

$$\frac{1}{2}\mu\mu'' + \frac{1}{2}\mu'\nabla^2 r - (\nabla^a \nabla^b r)(\nabla_a \nabla_b r) - \mu(\nabla^2 r)' = 0 \qquad (92)$$

respectively.

Eqns. (91) and (92) are still not ordinary differential equations: they contain derivatives of r. The final step is to evaluate $\nabla_a \nabla_b r$ for substitution into (91) and (92). We have:

$$t^b \nabla_a \nabla_b r = \nabla_a(t^b \nabla_b r) - (\nabla^b r)\nabla_a t_b = 0 - (\nabla^b r)[-\lambda^{-1} t_{[a} \nabla_{b]} \lambda]$$
$$= \frac{1}{2}\lambda^{-1} t_a(\nabla^b r)(\nabla_b \lambda) = \frac{1}{2}\lambda^{-1}\lambda' \mu t_a$$

and

$$(\nabla^b r)\nabla_a \nabla_b r = \frac{1}{2}\nabla_a(\nabla^b r \nabla_b r) = \frac{1}{2}\mu' \nabla_a r$$

Finally,

$$\underset{1}{l^a}\underset{1}{l^b}\nabla_a \nabla_b r = \underset{1}{l^a}\nabla_a(\underset{1}{l^b}\nabla_b r) - \underset{1}{l^a}(\nabla^b r)\nabla_a \underset{1}{l_b}$$
$$= 0 + \underset{1}{l^a}(\nabla^b r)\nabla_b \underset{1}{l_a} = \frac{1}{2}(\nabla^b r)[\nabla_b(\underset{1}{l^a}\underset{1}{l_a})]$$

and similarly for $\underset{3}{l^a}\underset{2}{l^b}$ and $\underset{3}{l^a}\underset{3}{l^b}$. Hence,

$$(\underset{1}{l^a}\underset{1}{l^b} + \underset{2}{l^a}\underset{2}{l^b} + \underset{3}{l^a}\underset{3}{l^b})\nabla_a \nabla_b r = \frac{1}{2}(\nabla^b r)\nabla_b(\underset{1}{l^a}\underset{1}{l_a} + \underset{2}{l^a}\underset{2}{l_a} + \underset{3}{l^a}\underset{3}{l_a})$$
$$= \frac{1}{2}(\nabla^b r)\nabla^b(2r^2) = 2r(\nabla^b r)(\nabla_b r) = 2r\mu$$

We now have expressions for the components of $\nabla_a \nabla_b r$ in the t^a-direction, the $\nabla^a r$-direction, and the $\underset{i}{l^a}\underset{i}{l^b}$ directions. Hence,

$$\nabla_a \nabla_b r = \frac{1}{2} \lambda^{-2} \lambda' \mu t_a t_b + \frac{1}{2} \mu^{-1} \mu' (\nabla_a r)(nabla\_br)$$
$$+ \mu r^{-1} [g_{ab} - \lambda^{-1} t_a t_b - \mu^{-1} (\nabla_a r)(\nabla_b r)] \tag{93}$$

Therefore,

$$\nabla^2 r = \frac{1}{2} \lambda^{-1} \lambda' \mu + \frac{1}{2} \mu' + 2\mu r^{-1}$$

$$(\nabla^a \nabla^b r)(\nabla_a \nabla_b r) = \frac{1}{4} \lambda^{-2} \mu^2 (\lambda')^2 + \frac{1}{4} (\mu')^2 + 2\mu^2 r^{-2}$$

Substituting, (91) and (92) become, respectively,

$$\lambda'' \mu - \frac{1}{2} \lambda^{-1} \mu (\lambda')^2 + \frac{1}{2} \lambda' \mu' + 2\mu \, r^{-1} \lambda' = 0 \tag{94}$$

$$-\frac{1}{4} \lambda^{-1} \mu \, \lambda' \mu' - \mu \, \mu' r^{-1} + \frac{1}{4} \lambda^{-2} \mu^2 (\lambda')^2 - \frac{1}{2} \lambda^{-1} \mu^2 \lambda'' = 0 \tag{95}$$

We have now obtained the ordinary differential equations we sought. What remains is to solve them. Eliminating λ'' between (94) and (95), we obtain simply $\lambda'/\lambda = \mu'/\mu$. So, λ is a constant multiple of μ. What multiple should we choose?

In Minkowski space, $\lambda = -1$, and $\mu = 1$, which suggests $\lambda = -\mu$. In particular, if we want things to be scaled properly asymptotically (far from the star, where space-time is nearly flat) we should make this choice. We do so. Setting $\lambda = -\mu$ in (95), we obtain immediately $\lambda''/\lambda' + 2r^{-1} = 0$. The solution is $\lambda = a + b/r$, where a and b are constants. What should we pick? Asymptotically, (i.e., for large r), we want λ to approach -1 (that is, we wish to so scale t^a). So, we choose $a = -1$. We write $\lambda = -1 + 2GM/r$ where G is the gravitational constant, and M is another constant. We wish to interpret M. Let us suppose we are far from the star, so $2Gm/r << 1$ (we shall discuss this issue in a moment). Then, to first order in $2GM/r$, $\frac{1}{2} \ln(-\lambda) = \frac{1}{2} \ln(1 - 2GM/r) = -GM/r$. But $\frac{1}{2} \ln(-\lambda)$ is, as we have seen, the quantity analogous to the Newtonian potential, while $-GM/r$ is the potential, in Newtonian theory, of a point mass of mass M. Hence, we interpret the number M as the "apparent mass of the star, as seen by its distinct gravitational effects". This remark is made more explicit by (85). Clearly, from this equation there follows an expression (which is not very interesting) for M as an integral (of $(\rho + p)$) over the star.

It should now be clear that one can choose coordinates in which the metric for the Schwarzschild solution takes the well-known form

$$-(1 - \frac{2GM}{r}) \, dt^2 + (1 - \frac{2GM}{r})^{-1} \, dr^2 + r^2(\, d\theta^2 + \sin^2 \theta \, d\varphi^2)$$

The θ and φ are "angular coordinates", while the scalar field r becomes a "radial coordinate".

Finally, we remark on the sign-reversal of $(1 - 2GM/r)$. when $r = r_c = 2GM$, this expression (for λ and μ) reverses sign. This r_c associated with any mass is called the *Schwarzschild radius* of that mass. (For the Sun, r_c is a mile or so; for the Earth, a millimeter.) Most bodies, of course, are larger then their Schwarzschild radius, so λ is always negative and μ positive. In collapse phenomenon, however, a body passes inside its Schwarzschild radius, to a region of space-time in which t^a becomes spacelike. Clearly, the approximation $2GM/r << 1$ is reasonable in most situations. Physically, r_c for a body is approximately the size that body must be in order that the Newtonian escape velocity equal the speed of light. Thus, "light cannot escape" from a body inside its Schwarzschild radius. This is reflected, geometrically, by a "turning in of the light cones", and, in the formalism, by the sign reversals of λ and μ.

26. The Schwarzschild Solution: Observations

Presumably, the Schwarzschild solution represents a good approximation to the geometry of space-time in the vicinity of our solar system. The fluid mass at the center is the Sun; the planets are "free test particles" which follow geodesics. Does this general relativistic model of our solar system agree in its broad features with the (very accurate) model from Newtonian gravitation? Are there any new, observable consequences of the model in general relativity? The answer to both questions is yes. The new features of the relativistic model are the three famous "tests" of general relativity.

We begin with the red-shift. The light we see from the surface of the sun is observed with a longer wavelength than the light had on leaving the surface. What is the prediction, from the Schwarzschild solution, of this effect? Of course the frequency (and wavelength) of a beam of light depends on a four-velocity of he who observes it. To discuss red-shift, therefore, we must decide who is looking at the light. Let us consider observers who are "static", i.e., whose four-velocities are proportional to the time-like Killing vector t^a. The four-velocity of such an observer is $\eta^a = (-\lambda)^{-1/2} t^a$ (since $t^a t_a = \lambda < 0$). Now consider a light ray with tangent vector k^a. Then the frequency of the light, as seen by our observer, is $\omega = -\eta^a k_a = -(-\lambda)^{-1/2} t^a k_a$. Let this light ray pass from point p to point q of space-time. Then, since $k_a t^a$ is constant along the geodesic, we have

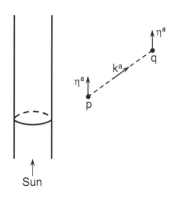

$$\frac{\omega(q)}{\omega(p)} = \left(\frac{\lambda(p)}{\lambda(q)}\right)^{1/2} = \left(\frac{1 - 2GM/r(p)}{1 - 2GM/r(q)}\right)^{1/2}$$

where we have used $\lambda = -(1 - 2GM/r)$. This is the formula for the red-shift. It enables us to compute the change in the frequency of light (as seen by an observer moving with the timelike Killing vector) as it passes from any point of space-time to any other point. In practice, one uses this equation in the case $2GM/r << 1$ (always such in the solar system, where $2GM$ is a kilometer). Then, to first order in $2GM/r$, the equation above becomes

$$\frac{\omega(q)}{\omega(p)} = 1 - \frac{GM}{r(p)} + \frac{GM}{r(q)} \qquad (GM << r)$$

In particular, for $r(q) > r(p)$, $\omega(q) < \omega(p)$, so the apparent wavelength of the light indeed increases as the light goes outward from the Sun. Solar red-shift experiments are good to about 10%, and agree with the theory. A much more accurate (tenths of a percent) red-shift experiment has been performed using the Earth, again, with, agreement with the theory.

The discussion of the other two tests requires more details about the behavior of geodesics. We now derive what we shall need. Consider a geodesic with tangent vector p^a, and fix a point q on this geodesic. Draw the two-sphere (of spherical symmetry) through q. Then we can choose the rotational Killing vectors so

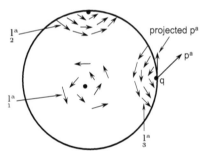

projected p^a

p^a

that $\underset{3}{l^a}$ vanishes at q, $\underset{1}{l^a}$ is parallel to the projection of p^a onto the sphere, and $\underset{2}{l^a}$ is orthogonal to the projection. (That such choices are possible is clear from the geometry of a sphere. Each Killing vector vanishes at a pair of antipodal points (the "axis of the rotation"). Choose $\underset{3}{l^a}$ to vanish at q, and $\underset{1}{l^a}$ and $\underset{2}{l^a}$ to vanish on the equator, where q is the North pole. Finally, let the projection of p^a into the sphere point along the meridian which goes to the point where $\underset{2}{l^a}$ vanishes.) At the point q, $p_a \underset{2}{l^a} = p_a \underset{3}{l^a} = 0$. Hence, this remains true along the entire geodesic. Therefore, it is true along the entire geodesic that $\underset{1}{l^a} l_a = r^2$. (At q, for example $\underset{1}{l^a} \underset{1}{l_a} = \underset{1}{l^a} \underset{1}{l_a}$, and $\underset{2}{l^a} \underset{2}{l_a}$, and $\underset{3}{l^a} \underset{3}{l_a} = 0$. Hence, from

(88) $\underset{1}{l^a} \underset{1}{l_a} = r^2$.)

From the choice above for Killing vectors, we have

$$p^a = \kappa t^a + \sigma \underset{1}{l}^a + \tau \nabla^a r \qquad (96)$$

where κ, σ, and τ are functions along the geodesic. To evaluate them, we use the fact that $E = -p_a t^a = -(\kappa t_a + \sigma \underset{1}{l}_a + \tau \nabla_a r) t^a = -\kappa\lambda$ is constant along the geodesic, and $L = p_a \underset{1}{l}^a = r^2 \sigma$ is constant along the geodesic. Hence,

$$p^a = -E\,\lambda^{-1} t^a + L\,r^{-2} \underset{1}{l}^a + \tau \nabla^a r \qquad (97)$$

where E and L are constants (the energy and angular momentum, respectively). To determine τ, we use the fact that $p^a p_a = -m^2$ is constant along the geodesic. Hence,

$$E^2 \lambda^{-1} + L^2 r^{-2} + \tau^2 \mu = -m^2 \qquad (98)$$

In (98), E, L and m are constants, while $-\lambda = \mu = (1 - 2GM/r)$. Hence, everything is known but τ, so (98) gives τ as a function of r. Thus, from symmetry considerations alone, we determine the tangent vector to our geodesic (97), given the values of the energy E, the angular momentum L, and the rest mass m.

The next step is to use (97) and (98) to determine the rate of change of certain quantities of interest along the geodesic. The first such quantity is r. We have $p^a \nabla_a r = (-E\,\lambda^{-1} t^a + L\,r^2 \underset{1}{l}^a + \tau \nabla^a r) \nabla_a r$. Since t^a and $\underset{1}{l}^a$ are both orthogonal to $\nabla_a r$ we have

$$p^a \nabla_a r = \tau\mu = \mu^{1/2} [-E^2 \lambda^{-1} - m^2 - L^2 r^{-2}]^{1/2} \qquad (99)$$

were, in the second step, we have used (98). The other "quantity of interest" is an angular function φ, defined by $\nabla^a r \nabla_a \varphi = t^a \nabla_a \varphi = 0$, $\underset{1}{l}^a \nabla_a \varphi = 1$ (i.e., defined by writing down the properties of the angular coordinate φ in Euclidean space). From (97),

$$p^a \nabla_a \varphi = L\,r^{-2} \qquad (100)$$

The equations we wanted are (99) and (100). Note, from (100), that φ is strictly increasing along the geodesic. Hence, we may take φ as a parameter along our geodesic, and take r along the geodesic as a function of this parameter φ. To find the differential equation for $r(\varphi)$, divide (99) by (100):

$$\frac{\mathrm{d}r}{\mathrm{d}\varphi} = r^2 L^{-1} \mu^{1/2} [-E^2 \lambda^{-1} - m^2 - L^2 r^{-2}]^{1/2} \tag{101}$$

$$= r^2 L^{-1} \left(\frac{1 - 2GM}{r} \right)^{1/2} \left[E^2 \left(1 - \frac{2GM}{r} \right)^{-1} - m^2 - L^2 r^{-2} \right]^{1/2}$$

The right side of (101) is an explicit function of r. Hence, (101) can, in principle, be integrated to give $r(\varphi)$. The problem of finding the geodesics in the Schwarzschild solution reduces to that of performing a certain integration. Unfortunately, the result is an elliptic integral, whose properties are not particularly transparent. We therefore proceed to analyze (101) in a more indirect way. First, replace r in (101) by $u = 1/r$, to obtain

$$\frac{\mathrm{d}u}{\mathrm{d}\varphi} = -L^{-1} [E^2 - m^2 (1 - 2GM\,u) - L^2\,u^2 (1 - 2GM\,u)]^{1/2}$$

Next, square this equation, and take $\mathrm{d}/\mathrm{d}\varphi$ of both sides:

$$2 \frac{\mathrm{d}u}{\mathrm{d}\varphi} \frac{\mathrm{d}^2 u}{\mathrm{d}\varphi^2} = \frac{2GM\,m^2}{L^2} \frac{\mathrm{d}u}{\mathrm{d}\varphi} - 2u \frac{\mathrm{d}u}{\mathrm{d}\varphi} + 6GM\,u^2 \frac{\mathrm{d}u}{\mathrm{d}\varphi}$$

Finally, rearrange terms:

$$\frac{\mathrm{d}^2 u}{\mathrm{d}\varphi^2} + u = \frac{GM\,m^2}{L^2} + 3GM\,u^2 \tag{102}$$

Clearly, every solution of (101) is also a solution of (102). It should also be clear that (102) is easier to think about than (101). Our approach to (102) will be based on an approximation procedure involving $GM/r << 1$, which, as we have seen, is well satisfied in our solar system.

We first consider the case $m > 0$, i.e., the case of orbits of material particles. To zeroth order in GM/r, we may ignore the term $3GM\,u^2$ on the right in (102), for it is negligible compared with u. Then (102) becomes $\frac{\mathrm{d}^2 u}{\mathrm{d}\varphi^2} + u = GM\,m^2/L^2$, with solution $u = GM\,m^2\,L^{-2} + GM\,m^2\,L^{-2}\epsilon \cos\varphi$, where ϵ is a constant. Expressed in terms, of r,

$$r = \frac{L^2}{GM\,m^2} \frac{1}{1 + \epsilon \cos\varphi} = r_0 \frac{1}{1 + \epsilon \cos\varphi} \tag{103}$$

Eqn. (103) will be recognized as the equation for an ellipse. Thus, we obtain, to zeroth order in GM/r, the usual elliptical Newtonian orbits. Note in particular that, to this order, $u(\varphi)$ is periodic in φ with period 2π.

The relativistic perturbations from the Newtonian results are obtained by considering the next order term in GM/r. We seek a solution of (102) of the form

$$u = GM\,m^2\,L^{-2}(1 + \epsilon\,\cos\varphi) + \delta u \tag{104}$$

where $\delta u << u$. Substituting, and keeping first order terms, we obtain

$$\frac{d^2}{d\varphi^2}\delta u + \delta u = 3\,G^3M^3m^4L^{-4}(1 + \epsilon\,\cos\varphi)^2 \tag{105}$$

This is the equation for the perturbation, δu from the Newtonian orbit. But (105) is easy to solve:

$$\delta u = 3\,G^3M^3m^4L^{-4}[1 + \frac{1}{2}\epsilon^2 + \frac{1}{6}\epsilon^2\,\cos 2\varphi + \epsilon\,\varphi\,\sin\varphi] \tag{106}$$

Hence, to first order in GM/r, the solution of (102) is

$$u = GM\,m^2\,L^{-2}(1 + \epsilon\,\cos\varphi)$$
$$+ 3\,G^3M^3m^4L^{-4}[1 + \frac{1}{2}\epsilon^2 + \frac{1}{6}\epsilon^2\,\cos 2\varphi + \epsilon\,\varphi\,\sin\varphi] \tag{107}$$

What is the effect of the perturbation term in (107)? The terms $1 + \frac{1}{2}\epsilon^2 + \frac{1}{6}\epsilon^2\,\cos 2\varphi$ are periodic in φ with period 2π. These terms change the "shape" of the orbit, but do not cause the ellipse to precess. The precession is caused by the term $\epsilon\,\varphi\,\sin\varphi$ on right in (107). To see its effect, let us ignore" the shape-changing terms", $1 + \frac{1}{2}\epsilon^2 + \frac{1}{6}\epsilon^2\,\cos 2\varphi$, on the right in (107) (these are unobservable.) Then (107) becomes

$$u = GM\,m^2\,L^{-2} + \epsilon\,GM\,m^2\,L^{-2}\,\cos\varphi + 3\,\epsilon\,G^3M^3m^4L^{-4}\varphi\,sin\varphi \tag{108}$$

But, for $\kappa << 1$, $\cos(1 + \kappa)\varphi = \cos(\kappa\varphi)\cos\varphi - \sin(\kappa\varphi)\sin\varphi = \cos\varphi - \kappa\varphi\sin\varphi$. Hence, to first order in GM/r, (108) gives

$$u = GM\,m^2\,L^{-2}[1 + \epsilon\,\cos(1 - 3\,G^2M^2\,m^2\,L^{-2})\varphi] \tag{109}$$

The situation should now be clear. The elliptical Newtonian orbit of the planet precessed with each revolution of the planet. For each orbit, the axis of the ellipse moves through angle

$$\Delta\varphi = 6\,\pi\,G^2 M^2\,m^2\,L^{-2} = 6\,\pi\,G\,M/r_0 \qquad (110)$$

where, in the second expression, we have set $r_0 = L^2/GM\,m^2$ (see (103)). Eqn. (110) is the equation for the precession of the perihelion of a planetary orbit, so the precession effect is first order in GM/r_0.

To summarize, general relativity causes "corrections" in the Newtonian planetary orbits. The orbit changes its shape (which is unobservable), and, in addition, the perihelion (point of closest approach of the planet to the Sun) precesses, for each orbit, through an angle given by (110). For the planet Mercury, such a precession has been observed (the angle is about 10^{-7} radians per orbit), and agrees with the relativistic prediction to within about 5%.

Finally, we consider the geodesic equation, (102), in the case $m = 0$. Thus, we consider the behavior of light rays. This yields the third classical test of general relativity. With $m = 0$, (102) becomes

$$\frac{\mathrm{d}^2 u}{\mathrm{d}\varphi^2} + u = 3\,G\,M\,u^2 \qquad (111)$$

Again, we solve (111) by an approximation procedure, using $GM/r \ll 1$. Ignoring the small right side of (111) (i.e., working to zeroth order in GM/r) we have the solution

$$u = \frac{1}{r_0}\cos\varphi \qquad (112)$$

This, of course, is the equation for a straight line in polar coordinates. The first order term in GM/r changes this curve slightly. To find the change, we set

$$u = \frac{1}{r_0}\cos\varphi + \delta u \qquad (113)$$

with $\delta u \ll u$. Substituting into (111), and keeping first order terms,

$$\frac{\mathrm{d}^2\delta u}{\mathrm{d}\varphi^2} + \delta u = \frac{3\,G\,M}{r_0{}^2}\cos^2\varphi \qquad (114)$$

The solution of (114) is $\delta u = (GM/r_0{}^2)(2 - \cos^2\varphi)$. Hence, the solution of (111), good to first order in GM/r, is

$$u = \frac{1}{r_0}\cos\varphi + \frac{2\,G\,M}{r_0{}^2} - \frac{G\,M}{r_0{}^2}\cos^2\varphi \qquad (115)$$

Here, we are concerned with the asymptotic behavior of the light ray, i.e., its behavior far from the star. But r approaches infinity when u ($= 1/r$) approaches zero. Hence, we are interested in the φ-values for which the right side of (115) vanishes. To first order in GM/r, the right side of (115) vanishes for $\cos\varphi = -2\,GM/r_0$, or for $\varphi = \frac{\pi}{2} + \frac{2GM}{r_0}$ and $-\frac{\pi}{2} - \frac{2GM}{r_0}$. Clearly, the light "bends" as it passes the Sun, through an angle

$$\Delta\varphi = \frac{4\,G\,M}{r_0} \qquad (116)$$

This is the equation for the deflection of light passing by a body in general relativity. Note that r_0 on the right in (116) is the distance of the light from the Sun at the point of closest approach.

To summarize, light passing by a body is bent by the presence of that body, through an angle given by (116). This effect has been observed, by looking at light from distant stars passing close to the Sun. The observations agree with general relativity to perhaps 25%.

In interpreting the last two tests, we have implicitly assumed that, when an effect was expressed in terms of r and φ, then this is the effect that should be observed, using Euclidean coordinates in our solar system. This assumption should, strictly speaking, be justified by a more detailed analysis of what is actually observed. Such a justification is possible, but rather tedious. Suffice it to say that, for effects to first order in GM/r, r and φ adequately describe the "Euclidean geometry" with respect to which observationalists would orient themselves to report the results of their experiments.

27. The Initial-Value Formulation

An essential concept in physics is that of prediction. Given sufficient detail about what is happening now, one predicts, using the equation which describe the system, what will happen later. This power of prediction is one of the important things which makes a physical theory a physical theory. It is available in essentially every physical theory. That prediction (of the future from the present) is indeed possible within the context of a given theory is made manifest by casting that theory into initial-value form. The initial-value formulation of a theory normally has the following appearance. There is a certain set of information about the system (at any one instant of time) called the data for the system. The theory gives a set of differential equations of the form $\mathrm{d}/\mathrm{d}t(\text{data}) = F(\text{data})$, where $\mathrm{d}/\mathrm{d}t$ is time derivative, and F is some function of the data. These are the evolution equations. If we know the data now, the evolution equations determine the data an instant later, then the data another instant later, etc. In this way, one predicts.

Of course, the notions above are not natural ones in general relativity. They refer to "time", and "things happening". In general relativity, there is only space-time: there is no time, no evolution – nothing happens. It is of interest, nonetheless, to ask whether general relativity can be forced into initial-value formulation. The answer is that, if one is willing to sacrifice some of the beauty, it can.

We wish to speak of time. Hence, we introduce a scalar field t on space-time with $\nabla_a t$ timelike. The various surfaces $t = \text{const}$ (one surface for each value of "const") thus represent surface of constant time. These are three-dimensional spacelike surfaces, and they fill space-time. Of course, a given space-time can be "sliced" as below in many different ways. This reflects the fact that "time" is not natural in general relativity. In order to introduce time, we must introduce additional structure (the t above) into space-time.

The initial-value formulation of general relativity results from recasting Einstein's equation in the presence of a scalar field t with timelike gradient. We first derive the equations, and then, at the end, discuss their signifi-cance.

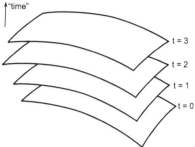

Set $\varphi = [-(\nabla^a t)(\nabla_a t)]^{-1/2}$, and $\xi^a = -\varphi \nabla^a t$. Thus, ξ^a is a unit ($\xi^a \xi_a = -1$) vector field in the direction of t. That is, ξ^a is the unit normal vector field to our family of surfaces. Furthermore, $\varphi \xi^a \nabla_a t = 1$. Thus, we may interpret $\varphi \xi^a$ as the "connecting vector" between nearby surfaces. If, for example, φ were constant, then all the surfaces would be equidistant from their (infinitesimally nearby) neighbors. For large φ, the surfaces are far apart; for small φ, close together.

A tensor field in space-time will be called *spatial* (with re-spect to the "field of observers" defined by ξ^a) if any index of that tensor field, contracted with ξ^a, gives zero. In particular, $h_{ab} = g_{ab} + \xi_a \xi_b$ is a spatial ten-sor field, the spatial metric. As

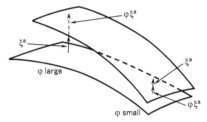

usual, $h_a{}^m h_{bm}$ so h_{ab} is also a spatial projection tensor. Finally, we introduce a dot for $\mathscr{L}_{\varphi \xi}$. Recalling that $\varphi \xi^a$ is the connecting vector between nearby surfaces, we can think of $\mathscr{L}_{\varphi \xi}$ as the "time derivative". We shall also need the spatial derivative operator, defined by

$$D_a T_{b...c} = h_a{}^m h_b{}^n \cdots h_c{}^p \nabla_m T_{n...p} \qquad (117)$$

on any spatial tensor field. That this is the correct spatial derivative can be verified by checking that $D_a h_{bc} = 0$ (the defining property of the derivative).

The idea is to introduce certain tensor fields on each surface $t =$ constant to represent the data. Then, we wish to obtain equations giving the "dot" of each data tensor field in terms of the values of those fields. In this way, we acquire the capability of integrating – evolving – from one surface to the next.

The first tensor field is the spatial metric h_{ab}. The second is a field Π called the extrinsic curvature, and defined by

$$\Pi^{ab} = \nabla^a \xi^b + \varphi^{-1} \xi^a \nabla^b \varphi + \varphi^{-1} \xi^a \xi^b (\xi^m \nabla_m \varphi) \qquad (118)$$

We first show that this Π^{ab} is a symmetric, spatial tensor field. To show symmetry, antisymmetrize (118) over "a" and "b":

$$\Pi^{[ab]} = \nabla^{[a}\xi^{b]} + \varphi^{-1}\xi^{[a}\nabla^{b]}\varphi = \varphi\nabla^{[a}(\varphi^{-1}\xi^{b]}) = 0 \qquad (119)$$

where, in second step, we have combined terms, and, in the third, we have used the fact that $\varphi^{-1}\xi_a = -\nabla_a t$ whence the curl of this vector field vanishes. To see that Π^{ab} is spatial, contract with ξ_b

$$\begin{aligned}
\xi_b\Pi^{ab} &= \xi_b\nabla^a\xi^b + \varphi^{-1}\xi^a\xi_b\nabla^b\varphi + \varphi^{-1}\xi_b\xi^a\xi^b(\xi^m\nabla_m\varphi) \\
&= 0 + \varphi^{-1}\xi^a(\xi^b\nabla_b\varphi) - \varphi^{-1}\xi^a(\xi^m\nabla_m\varphi) = 0
\end{aligned} \qquad (120)$$

Since $\Pi^{ab} = \Pi^{ba}$, it follows also that $\xi_a\Pi^{ab} = 0$.

The data for general relativity are the two tensor fields and h_{ab} and Π^{ab}. We think of h_{ab} as analogous to "x" (position of a particle) in particle mechanics, and Π^{ab} as analogous to p (momentum of the particle). If you are given the position and momentum of a particle initially, the future motion of the particle is determined. Similarly, if you are given the "configuration of the geometry" (the spatial metric h_{ab}) and its "momentum" (the extrinsic curvature Π^{ab}), the spacetime is determined. We now derive the equations which describe how this "determination" comes about.

We first consider h_{ab}:

$$\begin{aligned}
h'_{ab} &= \mathscr{L}_{\varphi\xi}h_{ab} = h_a{}^m h_b{}^n \mathscr{L}_{\varphi\xi}h_{mn} \\
&= h_a{}^m h_b{}^n[\varphi\,\xi^p\nabla_p h_{mn} + h_{mp}\nabla_n(\varphi\,\xi^p) + h_{pn}\nabla_m(\varphi\,\xi^p)] \\
&= h_a{}^m h_b{}^n[\varphi\,\xi^p\nabla_p(g_{mn} + \xi_m\xi_n) + \varphi\,h_{mp}\nabla_n\xi^p + \varphi\,h_{pn}\nabla_m\xi^p] \\
&= 2\,\varphi\,\Pi_{ab}
\end{aligned} \qquad (121)$$

where, in the second step, we have used $\varphi\,\xi^b\mathscr{L}_{\varphi\xi}h_{ab} = \mathscr{L}_{\varphi\xi}(\varphi\,\xi^b\,h_{ab}) - h_{ab}\mathscr{L}_{\varphi\xi}(\varphi\,\xi^b) = 0 - 0 = 0$, in the third step, we have used the definition of the Lie derivative, in the fourth step, we have used $\xi^p h_{mp} = 0$ and $\xi^p h_{pn}$, and, in the fifth step, we have noted that the first term on the right in brackets gives zero, while, since

$$\Pi_{ab} = h_a{}^m h_b{}^n \nabla_m\xi_n \qquad (122)$$

the last two terms each give $\varphi\,\Pi_{ab}$.

Eqn. (121) is analogous to $\dot{x} = (1/2m)p$ in particle mechanics. It expresses the derivative of the "position" (h_{ab}) in terms of the "momentum". Thus, (121) supports the interpretation of Π^{ab} as a "momentum of geometry".

We now obtain two additional equations relating Π^{ab} and h_{ab}. The interpretation follows the derivation. From (122),

$$
\begin{aligned}
D_a \Pi_{bc} &= h_a{}^m h_b{}^n h_c{}^p \nabla_m [h_n{}^r h_p{}^s \nabla_r \xi_s] \\
&= h_a{}^m h_b{}^n h_c{}^p [(\nabla_m h_n{}^r) h_p{}^s \nabla_r \xi_s \\
&\quad + h_n{}^r (\nabla_m h_p{}^s) \nabla_r \xi_s + h_n{}^r h_p{}^s \nabla_m \nabla_r \xi_s] \\
&= h_a{}^m h_b{}^n h_c{}^p [(\xi^r \nabla_m \xi_n) h_p{}^s \nabla_r \xi_s \\
&\quad + h_n{}^r (\xi^s \nabla_m \xi_p) \nabla_r \xi_s + h_n{}^r h_p{}^s \nabla_m \nabla_r \xi_s]
\end{aligned}
\tag{123}
$$

Where, in the second step, we have expanded using the Leibniz rule. Consider the third step above. For the first term, $\nabla_m h_n{}^r = \nabla_m (g_n{}^r + \xi_n \xi^r) = \xi_n \nabla_m \xi^r + \xi^r \nabla_n \xi_m$. But the ξ_n in $\xi_n \nabla_m \xi^r$ gets annihilated by $h_b{}^n$. Thus, we get the first term on the right in the last step above. The second term is identical. Next, note that, since $\xi^s \nabla_m \xi_s = 0$, the second term on the right in (123) vanishes. To get rid of the first term, we antisymmetrize (123) over "a" and "b", so that term vanishes since $\Pi_{[mn]} = 0$. Thus, (123) becomes

$$
D_{[a} \Pi_{b]} = \frac{1}{2} h_a{}^m h_b{}^n h_c{}^p R_{mnpq} \xi^q
\tag{124}
$$

Eqn. (124) is interesting, but it is not quite what we want: we would prefer the Ricci tensor to the Riemann tensor on the right. Hence, we contract (124) over "a" and "c"

$$
D^c \Pi_{bc} - D_b \Pi^c{}_c = h_b{}^n R_{nq} \xi^q = 8\pi G\, h_b{}^n \xi^q T_{nq}
\tag{125}
$$

where, in the second step, we have used Einstein's equation.

The second equation we want relates the spatial curvature, $\mathscr{R}_{abcd}$, to other quantities. Let k_c be an arbitrary spatial vector field. Then, by definition of the (spatial) Riemann tensor,

$$
\begin{aligned}
\frac{1}{2} \mathscr{R}_{abc}{}^d k_d &= D_{[a} D_{b]} k_c = h_{[a}{}^m h_{b]}{}^n h_c{}^p \nabla_m (h_n{}^r h_p{}^s \nabla_r k_s) \\
&= h_{[a}{}^m h_{b]}{}^n h_c{}^p [(\nabla_m h_n{}^r) h_p{}^s \nabla_r k_s \\
&\quad + h_n{}^r (\nabla_m h_p{}^s) \nabla_r k_s + h_n{}^r h_p{}^s \nabla_m \nabla_r k_s] \\
&= h_{[a}{}^m h_{b]}{}^n h_c{}^p [(\xi^r \nabla_m \xi_n) h_p{}^s \nabla_r k_s \\
&\quad + h_n{}^r (\xi^s \nabla_m \xi_p) \nabla_r k_s + h_n{}^r h_p{}^s \nabla_m \nabla_r k_s]
\end{aligned}
\tag{126}
$$

where the steps are exactly those used in (123). As before, $\Pi_{[mn]} = 0$ implies that the first term on the right in (126) vanishes. For the second term on the right, $\xi^s \nabla_r k_s = \nabla_r (\xi^s k_s) - k_s \nabla_r \xi^s = -k^s \nabla_r \xi^s$, since k is spatial. For the third term on the right, use the definition of the (space-time) Riemann tensor. Thus, (126) becomes:

$$\frac{1}{2}\mathscr{R}_{abc}{}^{d}\,k_d = -\Pi_{c[a}\Pi_{b]d}\,k^d + \frac{1}{2}h_a{}^m h_b{}^n h_c{}^p R_{mnpd}\,k^d \tag{127}$$

Since k_c (spatial) is arbitrary,

$$\mathscr{R}_{abcd} = -2\,\Pi_{c[a}\Pi_{b]d} + h_a{}^m h_b{}^n h_c{}^p h_d{}^q R_{mnpq} \tag{128}$$

Eqn. (128) gives the spatial Riemann tensor, but unfortunately, it involves the entire (space-time) Riemann tensor. To get to the Ricci tensor, we first contract (128) over "a" and "c"

$$\mathscr{R}_{bd} = -\Pi^c{}_c\Pi_{bd} + h_b{}^n h_d{}^q R_{nq} + R_{mbpd}\,\xi^m \xi^p + \Pi^c{}_b\Pi_{cd} \tag{129}$$

We still have space-time Riemann tensor, so we contract again;

$$\begin{aligned}\mathscr{R} &= -\Pi^c{}_c\Pi^d{}_d + \Pi^{cd}\Pi_{cd} + R + 2\,R_{mn}\,\xi^m \xi^n \\ &= -\Pi^c{}_c\Pi^d{}_d + \Pi^{cd}\Pi_{cd} + 8\,\pi\,G\,T_{mn}\,\xi^m \xi^n\end{aligned} \tag{130}$$

where, in the second step, we have used Einstein's equation.

We pause here to discuss the significance of (129) and (130). These two equations involve the data h_{ab} and Π^{ab} (and also the stress-energy), but they do not involve time derivatives. Thus, the data cannot be arbitrary tensor fields (on one of the spacelike three-dimensional surfaces): these fields must satisfy (125) and (130), These are called the constraint equations: they constrain, the acceptable choices for initial data.

We now derive the final equation for the initial-value formulation of general relativity. It is the evolution equation for Π^{ab}:

$$\begin{aligned}\dot{\Pi}^{ab} &= \mathscr{L}_{\varphi\xi}\Pi^{ab} = \mathscr{L}_{\varphi\xi}(h^{am}h^{bn}\nabla_m\xi_n) \\ &= 2(\mathscr{L}_{\varphi\xi}h^{am})h^{bn}\nabla_m\xi_n + h^{am}h^{bn}\mathscr{L}_{\varphi\xi}(\nabla_m\xi_n) \\ &= -4\,\varphi\,\Pi^{am}\Pi^b{}_m + h^{am}h^{bn}[\varphi\,\xi^p\nabla_p\nabla_m\xi_n \\ &\quad + (\nabla_p\xi^n)\nabla_m(\varphi\,\xi^p) + (\nabla_m\xi_p)\nabla_n(\varphi\,\xi^p)] \\ &= -4\,\varphi\,\Pi^{am}\Pi^b{}_m + h^{am}h^{bn}[\varphi\,\xi^p\nabla_m\nabla_p\xi_n \\ &\quad + \varphi\,\xi^p R_{pmnq}\,\xi^q + (\nabla_p\xi_n)(\varphi\,\nabla_m\xi^p + \xi^p\nabla_m\varphi) \\ &\quad + (\nabla_m\xi_p)(\varphi\,\nabla_n\xi^p + \xi^p\nabla_n\varphi)]\end{aligned} \tag{131}$$

where, in the fourth step, we have used $\mathscr{L}_{\varphi\xi}\, h^{ab} = -2\,\varphi\,\Pi^{ab}$ (see (121), noting that $h^{ab}h_{cb} = h^a{}_c$), (122), and the expression for the Lie derivative, and in the fifth step, we have used $\nabla_p\nabla_m\xi_n = \nabla_m\nabla_p\xi_n + R_{pmnq}\,\xi^q$. We now do various things to various terms on the right in the last expression in (131). The first term is fine. For the second term, note that

$$\xi^p\nabla_m\nabla_p\xi_n = \nabla_m(\xi^p\nabla_p\xi_n) - (\nabla_m\xi^p)\nabla_p\xi_n$$
$$= \nabla_m(-\varphi^{-1}D_n\varphi) - (\nabla_m\xi^p)\nabla_p\xi_n$$

where we have used (from (118) $\xi^p\nabla_p\xi_n = -\varphi^{-1}D_n\varphi$. The third term is fine. For the fifth term, use $\xi^p\nabla_p\xi_n = -\varphi^{-1D_n\varphi}$, and, for the seventh term, use $\xi^p\nabla_p\xi_n = 0$. Substituting all this, using (122) repeatedly to replace first derivatives of ξ_a by Π_{ab}, (113) becomes

$$\dot{\Pi}^{ab} = -3\,\varphi\,\Pi^{am}\Pi^b{}_m + \varphi\,D^b(-\varphi^{-1}D^a\varphi)$$
$$= -R^{ambn}\,\xi_m\xi_n\varphi - \varphi^{-1}D^a\varphi\,D^b\varphi \tag{132}$$

But (132) still contains the space-time Riemann tensor, while we want only Ricci tensor things. We eliminate the Ricci tensor term from (132) by substituting (129) (which gives the offending term in terms of initial data). Making this substitution (and rearranging terms a bit) we obtain, finally,

$$\dot{\Pi}^{ab} = -D^aD^b\varphi - 2\,\varphi\,\Pi^{am}\Pi^b{}_m - \varphi\,\Pi^c{}_c\Pi^{ab} + \varphi\,\mathscr{R}^{ab}$$
$$- 8\,\pi\,G\,\varphi\,h^{am}h^{bn}(T_{mn} - \frac{1}{2}T\,g_{mn}) \tag{133}$$

where we have used Einstin's equation.

We have now obtained all the equations for the initial-value formulation of general relativity. (You've probably noticed that we keep doing the same sort of calculating over and over. These techniques are very important, and the derivations are well worth study. Once one gets used to them, the calculations are also very easy.) What remains is to interpret.

Let us consider the source-free case: $T_{ab} = 0$. Then our equations (125), (130), (121), and (133) become, respectively

$$D_b(\Pi^{ab} - \Pi\, h^{ab}) = 0 \tag{134}$$

$$\mathscr{R} - \Pi^{mn}\Pi_{mn} + \Pi^2 = 0 \tag{135}$$

$$\dot{h}_{ab} = 2\,\varphi\,\Pi_{ab} \tag{136}$$

$$\dot{\Pi}^{ab} = -D^a D^b \varphi - 2\,\varphi\,\Pi^{am}\Pi^b_{\ m} - \varphi\,\Pi\,\Pi^{ab} + \varphi\,\mathscr{R}^{ab} \tag{137}$$

where we have set $\Pi = \Pi^c_{\ c}$. The first two are called the *constraint equations*, the last two *evolution equations*.

The constraint equations contain neither time-derivatives nor φ. They represent conditions which must be satisfied by our data (h_{ab}, Π^{ab}). The evolution equations give the time derivative of each of h_{ab}, Π^{ab} in terms of the values of these fields. These two equations involve φ, which, physically, represents the "rate at which the evolution proceeds." Thus, φ allows us to evolve (from a given spacelike three-surface) more quickly in some regions than in others. That is, φ represents the fact that our introduction of "time" is unnatural in general relativity.

Suppose we are given a three-dimensional manifold S with a symmetric tensor Π^{ab} on it and a positive-definite metric h_{ab}. This is an initial data set for general relativity. Suppose, further, that these fields satisfy the constraint equations. Choose any field φ on S. Then the evolution equations determine h_{ab} and Π^{ab} an instant dt later. Now pick a φ on this new surface, and again from (136), (137), find h_{ab}, Π^{ab} another instant later. Continuing, one obtains a family of a "stacked up" three-surfaces, with h_{ab} and Π^{ab} on each. In other words, one obtains a space-time. Our freedom to choose φ at each instant reflects the fact that, from an initial three-surface, there are many ways to draw successive surfaces. From these remarks, one would expect that the constraints are preserved in the evolution, i.e., that the "dots" of (134) and (135) reduce, using (136) and verification is two or three pages of calculations).

Two analogies are represented in the table below:

General Relativity	Particle	Electromagnetism	
h_{ab}	x	B	position
Π_{ab}	p	E	momentum
φ	–	–	"evolution rate"
$\nabla_b(\Pi^{ab} - \Pi h^{ab}) = 0$	–	$\nabla \cdot E = 0$	constraint
$\mathscr{R} - \Pi^{mn}\Pi_{mn} + \Pi^2 = 0$	–	$\nabla \cdot B = 0$	constraint
$\dot{h}_{ab} = 2\,\varphi\,\Pi_{ab}$	$\dot{x} = \frac{1}{2m}p$	$\dot{B} = -\nabla \times E$	evolution
$\dot{\Pi}^{ab} = -D^a D^b \varphi - 2\,\varphi\,\Pi^{am}\Pi^b_{\ m} - \varphi\,\Pi\Pi^{ab} + \varphi\,\mathscr{R}^{ab}$	$\dot{p} = -m\nabla V$	$\dot{E} = \nabla \times B$	evolution

There is nothing in either the case of a particle (in potential V) or in electromagnetism which corresponds to the "evolution rate". The particle and electromagnetic case have a "natural time", and so don't need a special function to represent the rate. There are no constraints for the particle in potential (although there would be if we required that the particle always remain in a certain surface in space). There are constraints in the Maxwell case. Everything has its evolution equations.

In principle, one could imagine discovering solutions of Einstein's equation by finding solutions of (134), (135), and integrating them into the future using (136), (137). In practice, this is not a very fruitful approach.

The existence of an initial-value formulation of general relativity is interesting, in my view, for two reasons: it provides insight into the structure of the theory, and because, in numerous situations in everyday life, one can make an argument or make it clearer by referring to these equations. An example of these reasons follows in the next section.

28. Signal Propagation

What is the speed of propagation of gravitational effect? First note that this is a question which can, in principle, be answered experimentally. Two observers station themselves on separate mountain peaks. One observer holds a rock in each hand at his side, and, at a certain instant, lifts the rock so his arms are outstretched at his sides. From the Newtonian point of view, he produces a mass distribution with a quadruple moment. The observer on the other mountain has sensitive instruments with which he can measure gravitational fields. This second observer records the instant at which his instrument first detect the rock motion. In this way, one could determine a quantity we might call the "speed of propagation of gravitational signals".

What prediction does general relativity make on this question? Let us first see how, in principle, one might calculate such a thing. One would look around for an exact solution of Einstein's equation in which the stress-energy would reasonably be interpreted as representing "the Earth, with two mountains, two observers, one with a rock in each hand, who separates the rocks, and, with the other observer, some sensitive instruments,..." One would then reproduce, within this space-time, the measurements the observers made when they decided what they meant by the "distance between the mountains", etc. It is clear that this is no way to attack the question.

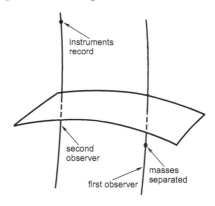

The initial-value formulation of general relativity offers a comparatively simple way of formulating this question. As an example of an application of this formulation, we now discuss the question of signal propagation.

Note, firstly, that, in our discussion above of a typical experiment, the notion of an initial-value formulation was implicit. The first observer's separating the masses "caused" the second observer's instruments to respond. It was implicitly assumed that what happened in the future (instrument response) was predicated on what had happened earlier (mass separation). It is perhaps not unreasonable, therefore, that the initial-value formulation is appropriate for the discussion.

Let us draw a space-time picture on our experiment. Each observer has a world-line, and we have the events "masses separated" and "instruments respond". We now draw a three-dimensional spacelike surface S in this space-time, where this surface cuts the first observer's world-line just above the event of mass separation. Let us now imagine two situations: the first observer either does or does not separate the masses on cue. Finally, we consider the surface S as a surface on which initial data for our space-time is specified. Now, the question of whether or not the masses were separated determines what the initial data on S is near where the first observer's world-line crosses S. In other words, we can imagine two initial data sets: that when the masses are and are not separated.

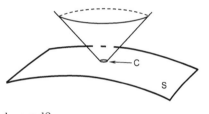

Similar experiments reduce to the same question. We are given initial data on a spacelike three-dimensional surface S. We choose a small region C on S. We change the data within C, leaving it alone outside C. In what region of space-time metric changed?

This question is essentially the question of signal propagation. The answer is: the space-time metric is altered (via the evolution equations, within the future light-cone from C. (The proof involves a rather complicated excursion into the properties of hyperbolic partial differential equations.) Thus, gravitational signals propagate at the speed of light (i.e., the light cone "spreads out" at the speed of light, just as gravitational effect do).

In this section, we have shown nothing. We have merely remarked that the initial-value formulation of general relativity leads to a precise and relatively straightforward expression of a question of physical interest. Although the mathematics is still not simple, at least we have the language to ask what we want to ask.

29. Time-Orientation

Let p be an event of the space-time manifold M. The two light-cones at p are qualitatively different physically. One calls one
the "future light cone", the other the "past
light-cone". One might, therefore, imagine
regarding only those space-times as reasonable which admit, globally, such a "past-
future distinction". We shall introduce a
bit of the mathematics involved in the discussion of such questions.

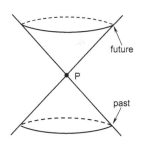

A space-time is said to be *time-orientable* if a continuous choice
of a future light-cone can be made globally, throughout space-time.
The remarks above suggest that our experiences in our region of the
Universe provide such a choice in our region of space-time. If, furthermore, we assume that, qualitatively, things are the same everywhere,
then we might expect reasonable space-times to be time-orientable.

We first give an example to
show that time-orientability is
not already a consequence of
what a space-time is. It is convenient, to make the drawing of

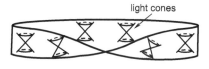

pictures easier, to consider two (rather the four) dimensions. Consider a Mobius strip with light-cones drawn on it as shown. Clearly, a
choice of one half of the full light-cone to be called "future" will lead to
problem if we try to maintain that choice consistently over the entire
space-time. There are never such problems locally.

Let us try to formulate the situation more mathematically. Let p
be a point of space-time, and let γ be a close curve, beginning and
ending at p. At p, make a choice of one of the two light-cones to
be called "future". Now carry this choice continuously around the
curve γ. It may happen that, on returning to p, (keeping track of
the "future light cone" all the way), we find that what we are calling

the future light-cone is, as we come back to p (to compare with our
original choice) the past light cone. If this were the situation, it would
not even be possible to assign "future" for the cones along this single
curve – much less over the entire space-time.

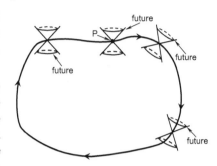

We write $L(\gamma) = +1$ if this
"light-cone reversal" does not
take place on passage around γ,
and -1 if it does.

We now claim that "failure
of time-orientability takes place
either around some closed curve
in space-time, or not at all in
space-time". More precisely, we
claim that a space-time is time-
orientable if and only if $L(\gamma) = +1$ for every closed curve γ. "Proof:"
It is clear that, if space-time is time-orintable, then $L(\gamma) = +1$ for every closed curve. Suppose, conversely, that $L(\gamma) = -1$ for every closed
curve. Pick any point p in space-time, and there choose one light-cone
to be called "future". Let q be any other point of space-time. Draw
a curve μ from p to q. Carrying our choice of "future light-cone"
continuously along this curve from p, we obtain a choice of a future
light-cone at q. Is this choice (at q) independent of the particle curve?
Let μ' be another curve from p to q. Then μ and μ' together, define
a closed curve beginning and ending at p. But L of this closed curve
is $+1$ (assumption). Hence, the designation of future at q is indeed
independent of the choice of curve. (This argument can easily be made
precise. It is not so easy for many later global arguments.)

Thus, failure of time-orientability is a
"curve property". Let's try to formulate that
more precisely. Fix a point p of M. Denote
by the collection of all closed curves which begin and end at p, where two closed curves such
that one can be continuously deformed into the
other are regarded as defining the same element of $\Pi_1(M)$. (More precisely: On the collection of all continuous closed curves from p
introduce the equivalence relation "are homo-
topic". Then, $\Pi_1(M)$ is the collection of equivalence classes.) For the cylinder, for example,
Π would be just the integers: the integer associated with a closed curve is the number of
times it goes around the cylinder before returning to its starting point.

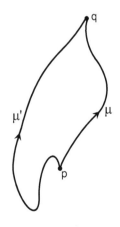

So far, this $\Pi_1(M)$ is just a set. Has it any other structure? We introduce a multiplicative structure. Let, γ_1 and γ_2 be two closed curves, each beginning and ending at p. We introduce a new closed curve, $\gamma_1\gamma_2$ as follows. This $\gamma_1\gamma_2$ is the curve obtained by first following γ_1 back to p, then following γ_2. It is clear that this product is associative: $\gamma_1(\gamma_2\gamma_3) = (\gamma_1\gamma_2)\gamma_3$. The unit curve is the unit which just remains at p – thus of course beginning and ending there. Finally, note that each curve has a "multiplicative inverse", namely the same curve described in the opposite direction.

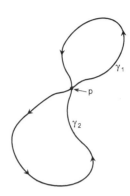

To summarize, $\Pi_1(M)$ has the structure of a group. It is called the "first homotopy group" of M. (It can actually be defined precisely. We have just given the intuitive flavor of things because the technicalities add little and use time which can perhaps be better spent elsewhere.)

We now make two observations. Firstly, $L(\gamma)$ is invariant under continuous deformation of γ (for L maps into a discrete set, $+1, -1$). Hence, L can be regarded as a mapping from $\Pi_1(M)$ to the set $(+1, -1)$. Secondly, we note that $L(\gamma_1\gamma_2) = L(\gamma_1)L(\gamma_2)$. In words, a time-reversing curve composed with a time-preserving curve is time-reversing, etc., for the other two possibilities. In other words, what we have shown is that L is a homomorphism from the group $\Pi_1(M)$ to the (multiplicative) group consisting of $(+1, -1)$. ("Homomorphism" means precisely $L(\gamma_1\gamma_2) = L(\gamma_1)L(\gamma_2)$.)

It is clear that the whole story regarding time-orientation (or lack of it) is summarized by this mapping L. In particular, a space-time M is time-orientable if and only if L maps all of $\Pi_1(M)$ to $+1$.

A space-time is said to be *simply connected* if $\Pi_1(M)$ consists of the identity alone (i.e., if every closed curve from p can be continuously deformed to the point p). For example, Minkowski space is simply connected, while a cylinder is not. Clearly, if $\Pi_1(M)$ consists only of the identity e, then $L(e) = +1$ (Proof: $L(ee) = L(e)L(e)$.) Thus, every simply connected space-time is time-orintable.

More, generally, note that the collection of elements of $\Pi_1(M)$ which L sends to $+1$ from a subgroup of $\Pi_1(M)$, and that this subgroup is of order two unless $L[\Pi_1(M)] = +1$. Thus, a space-time is time-orientable if $\Pi_1(M)$ has no subgroup of order two.

The remarks above are only intended to give a taste for how one attacks global questions. One begins (almost always!) with some notion of physical interest. One then tries to express these physical ideas in terms of more or less precise statements about the space-time.

Then, one is free to study such statements mathematically, for one has the assurance that results have at least the potential of having physical interest. We shall see a similar approach in each of the next sections.

30. Causality Violation

Consider a space-time M, g_{ab}, and suppose there exist within that space-time a closed timelike curve γ. Consider an observer who joins that curve at point p, and leaves it at q. Since the curve continues around from q to p, remaining timelike, our observer has the following opportunity. He could prearrange to have some rocket ship follow the closed curve from q back to p. Inside that ship, he could 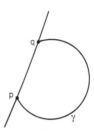 place a note informing himself to take some action between the events p and q. Thus, the possibility would be available for sending signals into one's own past: one could warn oneself of mistakes one has made in the past, in order to correct them.

Clearly the presence of closed timelike curves violates our aesthetic principles about how thins are put together. One might be tempted, on that ground alone, to require that no closed timelike curves exist. Only space-time in which this is the case would be regarded as "physically reasonable." But our experiences in physics have taught us to resist such temptations. Almost everything new and broadening in physics appeared, at first sight, to be impossible on aesthetic grounds.

Can more quantitative arguments be made suggesting that closed timelike curves should be ruled out? We make one suggestion in this direction. Suppose our Universe had a closed timelike curve passing through our region of space- 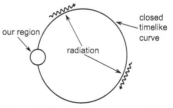 time. Suppose we, by means of magnets and charges, attempted to obtain a certain electromagnetic field (satisfying Maxwell's equation) in our region of space-time. Since we have closed timelike curves, the effects of such a field would, presumably, propagate around the curve, and re-influence our region of space-time. Thus, one might expect that, in the presence of closed timelike curves, electromagnetic fields must satisfy additional conditions than just Maxwell's equations. Since no such "additional conditions" are observed in Nature, the absence of closed timelike curves is suggested. Problem: Invent a precise, conclusive argument along these lines.

We have seen that the question of the existence of time-orientation is a "curve-question", in the sense that the situation can be adequately described in terms of $\Pi_1(M)$, the first homotopy group of M. Is some similar statement true concerning closed timelike curves? We show, by

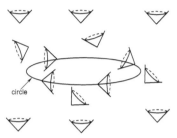

means of an example, that the answer is no. Our example is a simply connected space-time with closed space-time curves. (We consider three dimensions rather than four, in order to make it possible to draw a figure.) The example is shown in the figure above. In all regions except near a horizontal circle, the light-cones point upward as they usually do. As one approaches the circle, the cones begin to "tip" in the direction around the circle. On the circle itself, the cones have tipped so far that the circle is our closed timelike curve. But this space is simply connected.

Of course, in the example above, we have merely indicated a qualitative behavior for the light cones. We did not attempt to actually find a solution of Einstein's equation with these features. This is a standard technique in the study of global properties. If something can happen within space-time in general (ignoring Einstein's equation), then it can probably also happen for some solution of the equation. This suppression of the Einstein equation, of course makes things easier to think about. The phenomenon above, for example, actually occurs for certain solution, e.g., the Godel solution.

31. An Implication of Absence of Causality Violation

What sort of statements about the global structure of our Universe follow if one assumes absence of closed timelike curves? We give an example of one such statement in the present section.

Consider a space-time manifold M with metric g_{ab}. Let us suppose that there are spacelike surfaces S and S' in our space-time. (Physically, each represents *all of space at one instant of time.*) We suppose, furthermore, that the region of space-time between S and S' is compact. (A region of space-time is said to be compact if every sequence of points in that region, $P_1, P_2, \ldots$ has a sub-sequence, e.g., $P_7, P_19, P_237, \ldots$ which converges to some point of the region. A compact region of space-time is closed, and, intuitively, it does not "go off to infinity".)

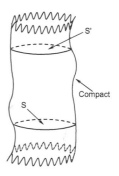

The closed (i.e., positive spatial curvature) Friedmann models provide an example of a space-time satisfying these conditions. For S and S', take the spatial sections. Then, since these sections are three-dimensional spheres, the region between two such sections is compact. Physically, we are considering cosmological models which are "spatially closed".

We shall show below that, if such a space-time is time orientable, and has no closed timelike curves, then S and S' are diffeomorphic (i.e., identical as manifolds). Physically, this means that time orientability and causality nonviolation (physically reasonable conditions) imply that the "topology of space" cannot change from one epoch to another in a closed universe. This result, in fact, does not require Einstein' equation: it just depends on a few geometrical properties of light cones.

Introduce an arbitrary positive-definite metric h_{ab} on M (Such always exists. This is a moderately subtle theorem about manifolds. The proof proceeds along the following lines: existence of g_{ab} on M implies that there is a derivative operator on M; existence of a derivative operator implies that M is paracompact; That M is paracompact implies existence of positive-definite metric.)

The next step in the construction is to obtain an everywhere timelike vector field on M. Fix a point p of M. Among all vectors ξ^a at p which are unit, future-directed, and timelike (with respect to g_{ab}), con-

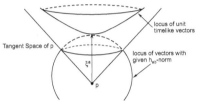

sider that one which has the smallest norm with respect to h_{ab}. This is our ξ^a. (Geometrically, the unit, future-directed timelike vectors at p describe a hyperbola in the tangent space. The vectors of a given norm with respect to h_{ab} describe a sphere. We adjust the radius of this sphere so that it intersects the hyperboloid in just one point.) Repeating at each point of our space-time, we obtain our desired timelike vector field on M. (This vector field has no direct physical significance. It merely provides a convenient way of keeping track of what the light cones are doing.)

Next, choose any point p of S, and draw the integral curve γ which begins at p. What can happen to this curve? It could return to S, remain within the region between S and S', or go to S'. The idea is to show that the first two possibilities cannot occur.

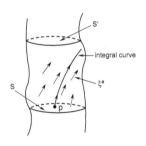

The curve γ cannot return to S, for our timelike vector field ξ^a points from S into the region between S and S'. We use compactness of this region to show that this possibility violates our assumption of no closed timelike curves. Choose a sequence of points $p_1, p_2, \ldots$ along the curve γ such that every point of γ lies between two points of this sequence. Here is a sequence on a compact region, so some subsequence approaches a point q of the region. Thus, the situation is that pictured on the right. Our curve continually sweeps by the point q, coming closer and closer, so a subsequence of our sequence approaches q. But now it is clear that we have a closed timelike curve in our space-time: merely distort γ on two of its successive passages near q to obtain a closed timelike curve, beginning and ending at q.

Thus, our curve cannot remain indefinitely in the region between S and S' without there existing closed timelike curves. Hence, γ goes to S'. But this is true for the curve beginning at each point of p and S. Thus, we obtain a mapping from S to S'. Repeating the same argument, from S' to S, we obtain an inverse mapping. This shows that S and S' are equivalent as manifolds. (More correctly, one would have to show that this mapping is smooth, with smooth inverse, to establish that it is a diffeomorphism.)

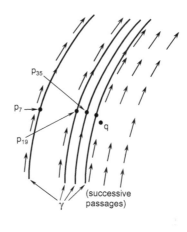

The argument above is intended merely as one example of how one proceeds to extract global information from an assumption of no casual anomalies. We repeat the conclusion in intuitive term: I"n a spatially closed, time-orientable, causally reasonable space-time, the topology of space cannot change from one epoch to another."

32. The Domain of Influence

In this section, we study the mathematical description of the question "Does what happen at event p influence what what happens at event q?", i.e., " Can a signal be sent from event p to event q?" This is clearly a fundamental relation between events. Again, our purpose is merely to give the flavor of how one describes things and what conclusion can be drawn.

Let M be a time-orientable space-time without closed timelike curves. Let p be a point of M. We denote by $I^-(p)$ the collection of all points q of the space-time which can be joined to p by a future-directed timelike curve. (Thus, $I^-(p)$ is what we have called the "interior of the past light-cone of p".) This $I^-(p)$ will be called the *past* of p, or the *past domain of influence* of p. (The latter term is perhaps unfortunate. $I^-(p)$ does not include its boundary, whereas points on the boundary can, in general, also influence p.) Physically, $I^-(p)$ represents the collection of all events of space-time which can affect what happens at p. Similarly, $I^+(p)$, the *future* of p, or *future domain of influence* of p, is the collection of all points which can be joined to p by a past-directed timelike curve.

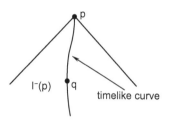

It is clear that $I^-(p)$ and $I^+(p)$ provide a concise formulation of the possibilities for events influencing other events. What remains is to describe some properties of $I^+(p)$ and $I^-(p)$.

We first remark that p is in $I^-(q)$ if and only if q is in $I^+(p)$. (That is, a point p can be joined to q by a future-directed timelike curve if and

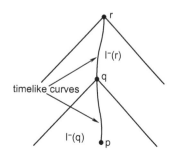

131

only if q can be joined to p by a past-directed timelike curve.) Furthermore, if p is in $I^-(q)$, and q is in $I^-(r)$, then p is in $I^-(r)$. (Proof: Join
the future-directed timelike curve from p to q to the future-directed
timelike curve from q to r. The result is a future-directed timelike
curve from p to r.)

We remark that $I^-(p)$ is always an open subset of space-
time. This fact is suggested by
the following argument. Firstly,
in Minkowski space, $I^-(p)$ is
open. Hence, if p is a point of
a general space-time, one would
expect that $I^-(p)$ is open in
some neighborhood of p. (For
locally, the metric is essentially
that of Minkowski space.) (Explicitly, one could introduce a chart in

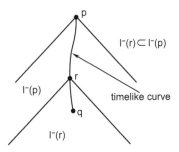

a neighborhood of p, explicitly write $I^-(p)$ in terms of coordinates,
and verify that it is open.) Thus, what remains is to show that this
"local openness of $I^-(p)$ implies global openness". Let q be a point of
our space-time in $I^-(p)$, so we have a future-directed timelike curve
from q to p. Choose a point r on this curve near q. Then, by construction, r is in $I^-(p)$, and q is in $I^-(r)$. But $I^-(r)$ is open near r,
so q is in this open set. But $I^-(r)$ is a subset of $I^-(p)$. Thus, we have
obtained an open subset of $I^-(p)$ which contains q. Since q was an
arbitrary point of $I^-(p)$, $I^-(p)$ is open.

The boundary of $I^-(p)$ is, roughly speaking, the "past null cone"
of p. One might expect, in particular, that it is a null surface. In fact,
this is true in a sense: $I^-(p)$ contains null geodesics. To prove this, let
q be a point of the boundary of $I^-(p)$. Choose a sequence of points
$p_1, p_2, \ldots$ of $I^-(p)$ which approach q. Since each of these points are
in $I^-(p)$, each can be joined to p by a future-directed timelike curve.

Let γ be the limit of this sequence of curves. Now, γ cannot be a timelike curve, for, if it
were, then q would lie in $I^-(p)$,
which, since $I^-(p)$ is open, would
contradict the fact that q is on
the boundary of $I^-(p)$. Thus, γ
must be a null curve. Furthermore, γ cannot enter $I^-(p)$, for

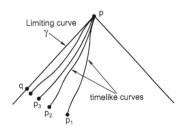

this, again would cause q to be in $I^-(p)$. Hence, γ must lie on the
boundary of $I^-(p)$. Next, note that γ cannot enter the future of q, for,
if it did so, one of the timelike curves from $p_1, p_2, \ldots$ must enter the

future of q (since $I^+(q)$ is open, and the sequence of timelike curves approaches γ). This implies that γ is a null geodesic. (A null curve from q in Minkowski space which is not a null geodesic enters the future of q. Thus, at any point at which γ is not a null geodesic, there γ would enter the future of q, by a local argument.)

To summarize, if q is in the boundary of $I^-(p)$, then, there is a future-directed null geodesic from q which lies in the boundary of $I^-(p)$. In this sense, then the boundary of $I^-(p)$ is a "null cone".

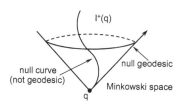

What can happens to this null geodesic? It may, as in Minkowski space, remain in the boundary of $I^-(p)$ until it reaches p. It is also possible that it never reach p. To construct an example of this behavior, consider Minkowski space with the closed region C removed from the manifold. In this example, the null geodesic from q meets C, and, since C has been removed, does not continue beyond this meeting. Thus, the null geodesic never reaches p.

To summarize, the structure suggested by the question "Can event p influence event q" is perhaps more fundamental than the manifold and metric structure which forms the basis for general relativity. This fact suggests that this causal structure

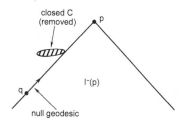

be studied in its own right. The mathematical description is in terms of $I^+(p)$ and $I^-(p)$, the future and past. We have defined these sets in a (time-orientable, no closed timelike curves) space-time, and obtained a few of their properties.

33. The Domain of Dependence

In Sect. 32, we discussed the question "Can event p influence event q?" We now consider the question "Does what happens on a surface S completely determines what happens at event q"? This leads to the notion of the domain of dependence. Again,

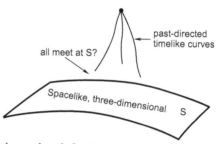

our purpose is merely to introduce the definition and obtain a few properties. We shall see that the domain of dependence is, in a sense, dual to the domain of influence.

Let M be a time-orientable space-time without closed timelike curves. Let S be a closed, three-dimensional, spacelike surface in M. Let p be a point to the future of S. Under what conditions would we expect what happens at p to be completely determined by information on S? Re-

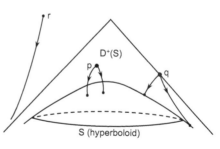

call that signals in general relativity propagate along timelike or null curves. Thus, we define the *future domain of dependence* of S, $D^+(S)$ as the collection of all points p of space-time such that every past-directed timelike curve from p meets S. (The definition above would make more sense physically if we replaced "timelike" by "timelike or null". It's only a question of whether or not $D^+(S)$ contains its boundary. For mathematical reasons, the above definition is more convenient.) Similarly, we define $D^-(S)$. (S will be included in $D^+(S)$ and

$D^-(S)$.) We give two examples of the domain of dependence. Let S be the surface $t = 0$ (in the usual coordinates) in Minkowski space. Then $D^+(S)$ is the region $t > 0$ in Minkowski space. For a less trivial example, let S be a hyperboloid in Minkowski space. Then $D^+(S)$ consists of all points between (and including) the hyperboloid and the light cone. Points p and q, for example, are in $D^+(S)$, for every past-directed timelike curve from either point reaches S. Point r is not in $D^+(S)$, because of past-directed timelike curve shown.

We now derive a few elementary properties of $D^+(S)$. The first is that $D^+(S)$ is closed. Let p be a point on the boundary of $D^+(S)$. Suppose that p were not in $D^+(S)$. Then there would be a past-directed timelike curve γ from p which fails to meet S. Choose a point r on this curve slightly to the past of p. Then r is is not in $D^+(S)$ (for

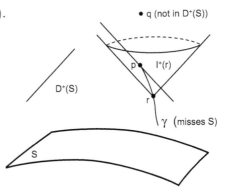

the rest of γ is a past-directed timelike curve from r which fails to meet S.) Hence, any point in the future of r is not in $D^+(S)$ (for, if q is in $I^+(r)$, then there is a past-directed timelike curve from q to r. Continue this curve along γ, so it misses S. We thus obtain a past-directed timelike curve from q which fails to meet S, so q is not in $D^+(S)$.) Now, $I^+(r)$ is open, contains p, and fails to meet $D^+(S)$. This contradicts our original placement of p on the boundary of $D^+(S)$. Hence, $D^+(S)$ is closed.

We next remark that, if p is in $D^+(S)$, and q is in $I^-(p)$, and in $I^+(s)$ for some point s of S, then q is in $D^+(S)$. Proof: suppose not. Then there is a past-directed timelike curve from q which misses S. Since q is in $I^-(p)$, there is a past-directed timelike curve from p which misses S. This contradicts our original placement of p in $D^+(S)$.

If follows immediately that, if p is in the boundary of $D^+(S)$, then $I^+(p)$ does not intersect $D^+(S)$. (Proof: If q is in $I^+(p)$ and $D^+(S)$, then p is in $I^-(q)$. By the remark above, all of $I^-(q)$ to the future of S is in $D^+(S)$. Hence, we have an open set in space-time which is in $D^+(S)$ and

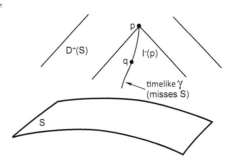

which includes p. This contradicts the placement of p on the boundary of $D^+(S)$.

As a final property of $D^+(S)$, we establish the following: If p is in the boundary of $D^+(S)$, then there is a past-directed null geodesic γ from p which remains in the boundary of $D^+(S)$. The proof is completely analogous to the proof that the boundary of the past of a point contains null geodesics. Choose a sequence of points, $p_1, p_2, \ldots$, which approach p, and which are not in $D^+(S)$. Then from each there is a past-directed timelike curve which fails to meet S. Take the limiting curve through p, and call it γ. Now, this γ must be timelike or null (since it is a limit of timelike curves). This γ cannot enter the interior of $D^+(S)$, for if it did, some timelike curve from some of $p_1, p_2, \ldots$ would enter $D^+(S)$, from which such a curve would necessarily meet S. Hence, γ cannot enter $I^-(p)$ to the future of S, for as we have seen, this open region is in $D^+(S)$. Thus, γ is a timelike or null curve which remains on the boundary of $I^-(p)$: hence γ is a null geodesic. Since γ never enters the interior of $D^+(S)$, and since γ remains of the boundary of $I^-(p)$ (which is in $D^+(S)$), γ must remain on the boundary of $D^+(S)$.

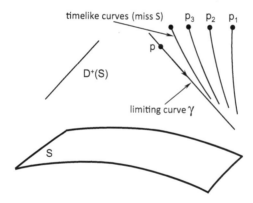

We conclude that, through every point of the boundary of $D^+(S)$, there passes a past-directed null geodesic which remains on the boundary of $D^+(S)$. See the first figure on page 162.

The domain of influence and the domain of dependence have many properties in common. This analogy is represented by the table below.

Concept	Domain of Influence	Domain of Dependence
Depends on	Point p $I^-(p)$	Spacelike Three-Dimensional Surface S $D^+(S)$
Figure		
Topology	Open	Closed
Boundary Point q	$I^-(q) \subset I^-(p)$ $I^+(q)$ not intersect $I^-(p)$	$I^-(q)$ (in future of S) $\subset D^+(S)$ $I^+(q)$ not intersect $D^+(S)$
Null Geodesics Through Boundary Point q	Future-directed, remains in boundary $I^-(p)$	Past-directed, remains in boundary $D^+(S)$

34. Singularity Theorems

In the last ten years or so, a number of results have been obtained or so, a number of results have been obtained to the effect that solutions of Einstein's equation have a tendency to become singular. We have already seen an example of singular behavior in space-time: in the Friedman solutions. There are rather analogous to the singular behavior one would see in Newtonian theory for an initially static ball of dust which is spherically symmetric – and which is allowed to collapse. All the dust particles fall toward the center, and there the density eventually becomes infinite. Why does nobody get excited about this Newtonian phenomenon? Because it is not "generic". If, for example, one endows our ball of dust with a small bit of angular momentum initially, then the ball will collapse a certain amount, but eventually the angular momentum will take over, and the ball will reexpand. The singularity in this Newtonian example arises only because of the special symmetry. Is a similar remark applicable to the singularity in cosmological (i.e., Friedman)solutions? This was for a long time a controversial question in general relativity. As we shall see in this section, the answer is no.

Let M, be a space-time, and let S be a three-dimensional, spacelike surface in M. Let p be a point of M to the future of S. We shall be concerned with the following question: Does there exist a timelike curve from p to S whose length in maximal? In 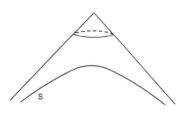 order to have an example before us, consider a hyperboloid S in Minkowski space. As we shall see shortly, it is the points between S and the light-cone (e.g., p in the figure, but not q) for which there does exist a longest timelike curve to S.

We return now to the general
case. Suppose, somehow, that
we did manage to find a timelike
curve from p to S whose length
is maximal. Then must be a

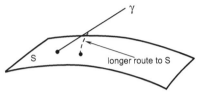

geodesic (for otherwise we could
lengthen by straightening it out; recall that a timelike curve which
"wiggles" gets near the light cone, and so tends to have a smaller
length), and must intersect S orthogonally (for otherwise we could
lengthen by moving slightly into point of intersection 121 Page with
S: again, this S is a statement which is obvious in Euclidean timelike curve from p to S – if one exists – is a geodesic which meets S
orthogonally. We are therefore led to study these normal geodesics.

We introduce the function φ
(defined, at least, near S) such
that $\varphi(p)$ is the distance from S
along the normal geodesic. Set
$\xi_a = \nabla_a \varphi$. (This vector field, although it will play an important
role, has little direct physical significance.) Then, since φ measures distance, we have $\xi^a \xi_a =$
-1. Note, furthermore, that
$\nabla_{[a}\xi_{b]} = \nabla_{[a}\nabla_{b]}\varphi = 0$. That
is, $\nabla_a \xi_a$ is symmetric. Hence,
$\xi^b \nabla_b \xi_c = \xi^b \nabla_c \xi_b = \frac{1}{2}\nabla_c(\xi^b \xi_b) =$

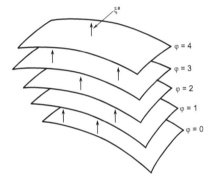

$\frac{1}{2}\nabla_c(-1) = 0$. Therefore, ξ^a is tangent to geodesics. Since, furthermore, ξ^a is normal to S (since $\varphi = 0$ on S), the integral curves of
ξ^a are precisely the normal geodesics from S – the candidates for the
longest timelike curves from points of M to S.

An important little calculation underlines all the singularity theorems. Set $c = -\nabla_a \xi^a$, the convergence of ξ^a. We are concerned with
the derivative of c along our geodesics:

$$\xi^m \nabla_m c = -\xi^m \nabla_m \nabla_a \xi^a = -\xi^m \nabla_a \nabla_m \xi^a - \xi^m R_{ma}{}^a{}_b \xi^b$$

$$= -\nabla_a(\xi^m \nabla_m \xi^a) + (\nabla_a \xi^m)(\nabla_m \xi^a) + \xi^m \xi^b R_{mb} \qquad (138)$$

$$= (\nabla^a \xi^b)(\nabla_b \xi_a) + \xi^a \xi^b R_{ab}$$

where, in the second step, we have used the definition of the Riemann
tensor; in the third step, we have differentiated by parts; and, in the
fourth step, we have used the definition of the Ricci tensor and the fact
that $\xi^m \nabla_m \xi^a = 0$. We next modify each term on the right in (138).
For the first term, note that $\nabla_a \xi_b$ is a symmetric, spatial tensor.

Write $\nabla_a \xi_b = -(1/3)c\,h_{ab} + (\nabla_a \xi_b + (1/3)c\,h_{ab})$, where $h_{ab} = g_{ab} + \xi_a \xi_b$. That is, we decompose $\nabla_a \xi_b$ into its "trace part" plus its "trace-free part". Then

$$(\nabla_a \xi_b)(\nabla^a \xi^b) = [-\frac{1}{3}c\,h_{ab} + (\nabla_a \xi_b + \frac{1}{3}c\,h_{ab})][-\frac{1}{3}c\,h^{ab} + (\nabla^a \xi^b + \frac{1}{3}c\,h^{ab})]$$

$$= \frac{1}{3}c^2 + (\nabla_a \xi_b + \frac{1}{3}c\,h_{ab})(\nabla^a \xi^b + \frac{1}{3}c\,h^{ab})$$

(139)

But the second term on the right in (139) is non-negative. (Proof: The trace of the square of a symmetric 3×3 matrix is non-negative.) Hence,

$$(\nabla_a \xi_b)(\nabla^a \xi^b) \geq \frac{1}{3}c^2$$

(140)

To treat the second term on the right in (138), we use Einstein's equation:

$$\xi^a \xi^b R_{ab} = 8\pi G(T_{ab} - \frac{1}{2}T g_{ab})\xi^a \xi^b$$

(141)

In the Newtonian limit, the right side in (141) is essentially the mass density (as seen by an observer with four-velocity ξ^a). Hence, this is positive in this limit. We now assume that the matter in our space-time is such that the right side of (141) is non-negative. This is called the *energycondition*. The energy condition is reasonable physically, firstly, because no matter has ever been observed which violates this condition, and secondly, because no theoretical models for matter have ever been constructed violating the energy condition. For a perfect fluid, for example the energy condition requires $\rho + p$ and $\rho + 3p$ be non-negative. For an electromagnetic field, the right side of (141) is $4\pi G(E^2 + B^2)$ (using E and B relative to ξ^a), and hence non-negative. Assuming that the energy condition is satisfied, (138) becomes,

$$\xi^m \nabla_m c \geq \frac{1}{3}c^2$$

(142)

But (142) can be solved (dividing by c^2). Let t be proper length along our geodesic, with $t = 0$ on S. Then the solution, $c(t)$, of (142) is (where $c_0 = c(0)$, the value on S)

$$c(t) \geq \frac{3}{3/c_0 - t}$$

(143)

Physically, (142) says that there is an irreversible tendency for our geodesics to converge – provided the energy condition is satisfied. The energy condition thus ensures that the matter acts "attractively", i.e.,

so as to pull the geodesics closer and closer together. This motion is
seen even more explicitly in (143). We see that c becomes infinite at
least by $t = 3/c_0$.

What does it mean geometrically
geometrically when c becomes infinite? How can the convergence
become infinite? It means that
our geodesics have begun to cross
(i.e., form a caustic). Of course,
this crossing of timelike geodesics
does not by itself indicate any singular behavior in our space-time.
Geodesics, for example, can cross in
Minkowski space. In the figure of

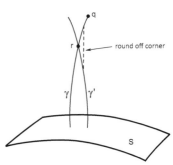

page 161, for example, the normal geodesics from S all meet at p –
but p is a perfectly regular point of space-time. We are interested in
this crossing of geodesics for a more subtle reason: because crossing
says something about the possibility of a geodesic's being the longest
curve to S. Suppose two nearby normal geodesics from S cross. Then
γ cannot be the longest timelike curve from q to S, for we can find a
longer curve by "rounding off the corner" at the crossing point r, and
then following γ' down to S.

We stop to summarize what we have so far. We have seen what,
if $c \geq c_0 > $ (c_0 a constant) on S, then every normal geodesic from S
meets another geodesic at least by the time it has gone a distance $3/c_0$.
(Here, the energy condition was required.) We have seen, furthermore,
that, if one normal geodesic crosses another normal geodesic, then the
first geodesic cannot be the longest timelike curve from a point beyond
the crossing point to S. Finally, we have seen that, if there is a longest
timelike curve from a point p to S, then that curve must be a geodesic
which meets S orthogonally. Taken together, these statements imply
that, *if we go farther than $3/c_0$ from S along any timelike curve, to
some point p, then there is no longest timelike curve from p to s.* (For,
if there were such a longest curve, it would be a normal geodesic with
length greater than $3/c_0$. But such cannot be the longest.)

The underline statement above is all we need to complete the proof.
It states that, sufficiently far away from S, the points of space-time
have the property that they cannot be joined to S by a longest timelike
curve.

We need one final assumption: that $D^+(S)$ is the entire future of
S. That is to say, we now assume that our space-time is such that
it is completely determined by what happens on S. (Such an S is
called a *Cauchy surface*.) The assumption that S is a Cauchy surface

implies that, from any p in $D^+(S)$, there is a longest timelike curve to S. This, again is a rather complicated technical result, but at least we can indicate why it is true. If p is in $D^+(S)$, then every past-directed timelike curve from p meets S. It can be shown from this that the closure of the union of these curves is compact. (Intuitively, this region cannot "go off to infinity," or else some timelike curve from p would "go off to infinity" missing S.) One then shows that the collection of all timelike or null curves from p to S is compact (in a suitable topology), and that length is a continuous function on this compact space of curves. Hence, "length" achieves its maximum.

We are now essentially done with the proof. If we go a distance $3/c_0$ from S along any timelike curve, we reach a point from which there is no longest timelike curve to S. But, if we assume that S is a Cauchy surface, then from every point, there is 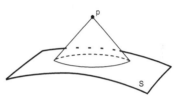 a longest timelike curve to S. But, if we assume that S is a Cauchy surface, then, from every point, there is a longest timelike curve to S. How can these statements be consistent? Only if one simply cannot go a distance more than $3/c_0$ from S along any timelike curve. Thus:

Let a space-time satisfy Einstein's equation, with a stress-energy satisfying the energy condition. Let that space-time have a Cauchy surface S on which $c \geq c_0$. where c_0 is a positive constant. Then no future-directed timelike curve from S has length greater than $3/c_0$.

There are two issues about this result which require discussion: What has it to do with our Universe? and what has it to do with singularities?

We have remarked that the Friedmann solutions appear to represent a good approximation to our Universe. Let S be one of the spacelike section in such a model. Then S is indeed a Cauchy surface, but the value of c on this surface is $-\alpha$ (where α is the Hubble function). Thus, c is not positive in the present epoch. The idea is to apply the result above, but with the roles of past and future interchanged. Thus, the convergence into the past is $\alpha > 0$, and all the conditions are satisfied. Thus, the result above states that timelike curves into the past cannot have length than $3/c_0$, and, indeed, in our Friedmann solutions, timelike curves into the past ran into the singularity, when ρ becomes infinite. The advantage of the result on the previous page over the explicit remark about the Friedmann solutions is that the result makes no assumptions about symmetric or other details of the solution – it is applicable to all solutions which display certain broad features. Thus, if we perturb (slightly) a Friedmann model (our

Universe, of course, would be a perturbed Friedmann solution, e.g., because isotropy does not hold exactly) we still must obtain the result that timelike curves into the past are no longer than a certain amount.

What has the result to do with singularities? The problem here is that we have required from the beginning that space-time be a smooth manifold with a smooth metric , and other smooth tensor fields. Thus, there is no way to represent singularities directly on a space-time manifold. (To give up smoothness leads to serious difficulties of principle.) If there were "actual singular points' 'on the space-time manifold, then such points would have to be removed to obtain the required smoothness. How would this "removal" be detected, given only the smooth space-time (after removal of all "singular points")? It could be detected by noting that certain timelike curves (namely, those which would have passed through the singular region) have only finite length in space-time. Thus, to describe singular behavior in space-time, one looks, not for certain quantities becoming infinite, but rather for timelike curves of finite length. Physically, av observer who follows such a curve has a finite total lifespan – "after that" he is no longer representable by a point on the space-time manifold. From the point of view of this observer, something singular is certainly going on.

The underline statement on the preceding page is rather weak. The problem lies with the assumption that S is a Cauchy surface. Suppose we are given initial data on S, and that we evolve this data to obtain a solution of Einstein's equation. It may just turn out that S will not be a Cauchy surface – that our Universe is such that it is not completely determined by what goes on on a certain spacelike three-dimensional surface. Consider, for example, the hyperboloid on page 165. The future domain of dependence for this S is the region between S and the light cone. It is not a Cauchy surface. Quite generally, singularity theorems which assume a Cauchy surface are rather weaker than those which do not. We now indicate how our theorem can be modified to eliminate the Cauchy surface assumption.

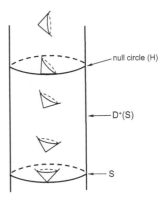

Assume that S is compact. Denote by H the boundary of the domain of dependence of S (S itself excluded).The crucial question is; Is this H also compact? Either answer leads to a difficulty.

Suppose first that H is compact. A two-dimensional space of this type is shown in the figure. The space-time is a cylinder, with the light-cones invariant under rotations about the axis. Near the bottom of the cylinder, the light-cones point up the cylinder. Further up, the light cones begin to "tip", so that, at a certain point, a circular cross-section of the cylinder is a null curve. After that, the cones continue to tip, giving closed timelike curves. the domain of dependence of S is the region between S and the null circle, while the null circle is H. This H is compact. The difficulty in this example is that we have closed timelike curves. We now show that a similar phenomenon occurs in the general case. We have seen in Sect. 33 that there are null geodesics which lie in H. Since H is compact, such a null geodesic must wander around in H until it eventually comes back arbitrarily close to itself. We have almost-closed null curves – a type of causality violation as serious, for all practical purposes as closed timelike curves. (There is an exact solution of Einstein's equation – called Taub-NUT space – in which precisely this phenomenon occurs.)

Now suppose that H is not compact. Then we can find a sequence of points, $p_1, p_2 \ldots$ in$D^+(S)$ which do not have a point of accumulation. Since each point is in $D^+(S)$ – where our earlier argument applies – from each point there is a past-directed timelike geodesic which meets S orthogonally, and which has length no greater than $3/c_0$.

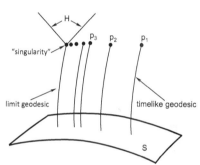

Take the limit of this sequence of geodesics. The result in another timelike geodesic. Clearly, the length of this geodesic cannot be greater than $3/c_0$, for, if it were, it would provide a point of accumulation for $p_1, p_2 \ldots$. Thus, we obtain a timelike geodesic from S of length no greater than $3/c_0$.

To summarize, we have

Let a space-time satisfy Einstein's equation, with a stress-energy satisfying the energy condition. Let that space-time have a three-dimensional spacelike surface S on which $c \geq c_0$, where c_0 is a positive constant. Let there be no almost closed null curves. Then some future-directed timelike geodesic from S has length no greater than $3/c_0$.

The first result requires a Cauchy surface, while the second only requires a compact spacelike surface. (That is to say, the second is applicable only to "closed universes.") The second result requires a causality condition. Finally, the first result gives a finite length for every timelike curve from S, while the second gives a finite length for just one timelike geodesic from S.

35. Conformal Transformation

In this section, we introduce a mathematical tool which finds numerous applications in general relativity.

Let M, g_{ab} be a space-time. A *conformal transformation* on g_{ab} consists of replacing g_{ab} by the metric $\tilde{g}_{ab} = \Omega^2$ where Ω is a positive scalar field on M. We call Ω the *conformal factor*. Geometrically, a conformal transformation amounts to "rescaling distances without changing angles". it is normally the case in practice that the new metric $\tilde{g}_{ab}$ has no direct physical significance. (It is, in fact, sometimes called the "unphysical metric".) A conformal transformation in merely a mathematical convenience.

Let, $T^{a \dots c}{}_{b \dots d}$ be some tensor field on M. Of course, the notion of a tensor field makes no direct reference to a metric. Hence, we still have this tensor field $T^{a \dots c}{}_{b \dots d}$ on M after some conformal transformation. It is convenient, however, to allow ourselves the following option. For a given conformal transformation $\tilde{g}_{ab} = \Omega^2$, we set $\tilde{T}^{a \dots c}{}_{b \dots d} = \Omega^n T^{a \dots c}{}_{b \dots d}$ where n is some real number (in practice, usually an integer). That is, we allow other tensor fields to be "rescaled" along with the metric. The number $m = n-$ (number of covariant indices on T) + (number of contravariant indices on T) is called the *dimension* of $T^{a \dots c}{}_{b \dots d}$. (Often, the term dimension refers to sec^m rather than the number m of itself.) For example, the metric g_{ab}, considered as just another tensor field, has dimension 0 $(= 2 - 2 + 0)$.

Suppose we have selected a dimension for $T^{a \dots c}{}_{b \dots d}$ and $S^{p \dots q}{}_{r \dots s}$, (i.e., suppose we have decided what $\tilde{T}^{a \dots c}{}_{b \dots d}$ and $\tilde{S}^{p \dots q}{}_{r \dots s}$ will be under a conformal transformation). What dimension should one choose for the tensor field $T^{a \dots c}{}_{b \dots d} \, S^{p \dots q}{}_{r \dots s}$. The natural choice is to assign to this outer product a dimension equal to the sum of the dimensions of $T^{a \dots c}{}_{b \dots d}$ and $S^{p \dots q}{}_{r \dots s}$. We shall always make this choice. The consequence is that the result of first taking the outer product and then applying tilde is the same as that of first applying tilde to both factors and then taking the outer product. We shall not (because we do not have to) make a separate decision about the assignment of dimension every time we take an outer product.

Again, let $T^{a\cdots c}{}_{b\ldots d}$ be a tensor field on M with some given dimension. What dimension should we assign to $T^{a.m.c}{}_{b.m.d}$ (one contraction)? A natural choice is to assign to $T^{a.m.c}{}_{b.m.d}$ the same dimension as was assigned to $T^{a\cdots c}{}_{b\ldots d}$. We shall always make this choice. The consequence is that the result of first contracting and then applying tilde is the same as that of first applying tilde and then contracting (for contracting annihilates one covariant and one contravariant index). We shall not (because we do not have to) make a separate assignment of dimension every time we contract over indices.

Once again, let $T^{a\cdots c}{}_{b\ldots d}$ be a tensor field on M with some given dimensions. We shall always assign to $T^{a\cdots c}{}_{b\ldots d}$ the same dimension as was assigned to $T^{a\cdots c}{}_{b\ldots d} = g_{ab}T^{a..q..c}{}_{b\ldots d}$. (Similarly, we assign to $T^{a\cdots c}{}_{b..}{}^{p}{}_{..d}$ the same dimension as was assign to $T^{a\cdots c}{}_{b\ldots d}$.) Thus, the result of lowering an index of a tensor field (using g_{ab}) and then applying tilde is the same as that of first applying tilde and then lowering the index (using $\tilde{g}_{ab}$). Proof: Lowering an index gives one more covariant index. But $\tilde{g}_{ab} = \Omega^2 g_{ab}$.) Of course, we set $\tilde{g}^{ab} = \Omega^{-2}g^{ab}$, so $\tilde{g}^{ab}\tilde{g}_{bc} = \tilde{g}^{a}{}_{c}\ (= \tilde{g}^{a}{}_{c})$. Thus, raising an index (with g^{ab}) and then applying tilde gives the same result as first applying tilde and then raising the index (with $\tilde{g}^{ab}$).

Finally, we adopt the convention that tensor fields with a tilde have their indices raised and lowered with $\tilde{g}^{ab}$ and $\tilde{g}_{ab}$, while tensor fields without a tilde have their indices raised and lowered (as always) with g^{ab} and g_{ab}. Now everything is consistent. One must make a choice, for each new tensor field in the space-time, of what its dimension will be. (This choice is based on convenience.) Dimensions add under outer product, and are unchanged under contraction and raising and lowering of indices. The metric itself has dimension zero (which, of course, it must have in order that the dimension remain invariant under raising and lowering of indices). (Note that $\tilde{g}^{ab}$ also has dimension zero.)

It should be emphasized that this scheme is natural and simple. The only freedom we had was the "2" in $\tilde{g}_{ab} = \Omega^2 g_{ab}$, and this choice was based on making formulae as simple as possible.

What dimension should we assign to the alternating tensor, ϵ_{abcd}? Clearly, it should have dimension zero (i.e., $\tilde{\epsilon}_{abcd} = \Omega^4 \epsilon_{abcd}$, for then we shall have $\tilde{\epsilon}_{abcd}\tilde{\epsilon}^{abcd} = -24$. That is, $\tilde{\epsilon}_{abcd}$ will be the alternating tensor for the metric $\tilde{g}_{ab}$.

We are now half done. We have treated the algebraic part of conformal transformations. What remains is the differential part. This requires, not more conventions, but rather a few calculations.

Let ∇_a and $\widetilde{\nabla}_a$ be the derivative operators with respect to g_{ab} and $\tilde{g}_{ab}$, respectively (so $\nabla_a g_{bc} = \widetilde{\nabla}_a \tilde{g}_{bc} = 0$). Then, since these are both derivative operators, there exists a tensor field $C^m{}_{ab} = C^m{}_{(ab)}$ on M such that

$$\widetilde{\nabla}_p T^{a...c}{}_{b...d} = \nabla_p T^{a...c}{}_{b...d} + C^a{}_{pm} T^{m...c}{}_{b...d} + \cdots +$$
$$+ C^c{}_{pm} T^{a...m}{}_{b...d} - C^m{}_{pb} T^{a...c}{}_{m...d} - \cdots - C^m{}_{pd} T^{a...c}{}_{b...m} \tag{144}$$

for every tensor field $T^{a...c}{}_{b...d}$ on M. Thus, this $C^m{}_{ab}$ gives us $\widetilde{\nabla}_a$ in terms of ∇_a. We wish to obtain an explicit expression for $C^m{}_{ab}$. We proceed as follows:

$$\begin{aligned} 0 = \widetilde{\nabla}_a \tilde{g}_{bc} &= \nabla_a \tilde{g}_{bc} - C^m{}_{ab} \tilde{g}_{mc} - C^m{}_{ac} \tilde{g}_{bm} \\ &= \nabla_a(\Omega^2 g_{bc}) - 2\Omega^2 C^m{}_{a(b} g_{c)m} \\ &= 2\Omega g_{bc} \nabla_a \Omega - 2\Omega^2 C^m{}_{a(b} g_{c)m} \end{aligned} \tag{145}$$

where, in the second step, we have used (144). But this equation is easy to solve:

$$C^m{}_{ab} = \frac{1}{2}\Omega^{-1} g^{mn}(g_{na}\nabla_b\Omega + g_{nb}\nabla_a\Omega - g_{ab}\nabla_n\Omega) \tag{146}$$

Thus, with the aid of (144) and (146), we can convert from $\widetilde{\nabla}$'s to ∇'s at will, obtaining, by such a conversion, additional terms involving derivatives of Ω.

We give two examples of these remarks. Let T^{ab} be a tensor field of dimension $\sec^{-24}$. Then

$$\begin{aligned} \widetilde{\nabla}_c \widetilde{T}^{ab} &= \nabla_c \widetilde{T}^{ab} + C^a{}_{cm} \widetilde{T}^{mb} + C^b{}_{cm} \widetilde{T}^{am} \\ &= \nabla_c(\Omega^{-6}T^{ab}) + \Omega^{-6}T^{mb}\tfrac{1}{2}\Omega^{-1}(\delta^a{}_c \nabla_m\Omega + \delta^a{}_m \nabla_c\Omega - g_{mc}\nabla^a\Omega) \\ &\quad + \Omega^{-6}T^{am}\tfrac{1}{2}\Omega^{-1}(\delta^b{}_c \nabla_m\Omega + \delta^b{}_m \nabla_c\Omega - g_{mc}\nabla^b\Omega) \\ &= \Omega^{-6}\nabla_c T^{ab} - 5\Omega^{-7}T^{ab}\nabla_c\Omega + \tfrac{1}{2}\Omega^{-7}\delta^a{}_c T^{mb}\nabla_m\Omega \\ &\quad + \tfrac{1}{2}\Omega^{-7}\delta^b{}_c T^{am}\nabla_m\Omega - \tfrac{1}{2}\Omega^{-7}T_c{}^b\nabla^a\Omega - \tfrac{1}{2}T^a{}_c\nabla^b\Omega \end{aligned}$$

where, in the first step, we have used (144), and, in the second, we have used (146).

As a second example, we evaluate the Riemann tensor, $\widetilde{R}_{abc}{}^d$, of the metric g_{ab}. As always, one obtains the Riemann tensor by commuting derivatives. Fix a vector field k_c on M. Then

$$\tfrac{1}{2}\widetilde{R}_{abc}{}^{d}\,k_d = \widetilde{\nabla}_{[a}\widetilde{\nabla}_{b]}k_c = \nabla_{[a}\widetilde{\nabla}_{b]}k_c - C^{m}{}_{[ab]}\,\widetilde{\nabla}_{m}k_c - C^{m}{}_{c[a}\,\widetilde{\nabla}_{b]}k_m$$

$$= \nabla_{[a}\widetilde{\nabla}_{b]}k_c - C^{m}{}_{c[a}\,\widetilde{\nabla}_{b]}k_m = \nabla_{[a}(\nabla_{b]}k_c - C^{m}{}_{b]c}\,k_m)$$

$$- C^{m}{}_{c[a}\,(\nabla_{b]}k_m - C^{n}{}_{b]m}\,k_n) = \nabla_{[a}\nabla_{b]}k_c - k_m\nabla_{[a}C^{m}{}_{b]c}$$

$$- C^{m}{}_{c[b}\,\nabla_{a]}k_m - C^{m}{}_{c[a}\,\nabla_{b]}k_m + C^{m}{}_{c[a}\,C^{n}{}_{b]m}\,k_n$$

$$= \tfrac{1}{2}R_{abc}{}^{d}\,k_d - k_d\nabla_{[a}C^{d}{}_{b]c} + C^{m}{}_{c[a}\,C^{d}{}_{b]m}\,k_d$$

where, in the second step, we have used (144), in the third step, we have used $C^{m}{}_{[ab]} = 0$, in the fourth step, we have again used (144), in the fifth step, we have expanded, and, in the sixth step, we have used the definition of the Riemann tensor. Since k_d is arbitrary, we have

$$\widetilde{R}_{abc}{}^{d} = R_{abc}{}^{d} - 2\nabla_{[a}C^{d}{}_{b]c} + C^{m}{}_{c[a}\,C^{d}{}_{b]m}$$

One could now substitute (146) and obtain a detailed expression for $\widetilde{R}_{abc}{}^{d}$ in terms of $R_{abc}{}^{d}$ and derivatives of the conformal factor.

It should be clear from the remarks above that, once a dimension has been assigned to each tensor field in sight, every tensor equation can be converted to a tilde-equation, and, conversely, tensor equations with tildes can be converted to equation without tildes.

By using conformal transformation, equations can sometimes be simplified or made more transparent geometrically. A conformal transformation is essentially the only simple thing one can do to a metric. Finally, conformal transformation can be used to bring out structure not apparent from the physical metric. We shall see an example of the latter application in the following sections.

36. Asymptotic Structure: Introduction

It is often the case in physics that one seeks a description of a system in which certain details of its structure are suppressed, and certain overall properties use in the description. (Example: The description of a gas by its pressure, density, and temperature.) Such a global description is possible, in special relativity, essentially because of the action of the Poincaré group as the group of isometries on Minkowski space. In particular, as we have seen, the presence of Killing vectors (the generators of Poincaré group – the "infinitesimal isometries") leads to the notions of the energy, momentum, angular momentum of a closed system in special relativity. Stated in somewhat more vague terms, the symmetry of Minkowski space allows a "comparison of tensors at different space-time points". Hence, one can "add effects from various regions of space-time to obtain a description of our system as a whole".

Unfortunately, in general relativity there are, in the "generic case", no Killing vectors at all – no symmetries. Thus, the special relativistic possibilities for an overall description of a closed system are lost. Can anything at all be done in this direction in curved space? Does there exist an – even partial – global description of a closed system in general relativity?

Such a description is indeed possible if we restrict consideration to space-time which are asymptotically flat. Asymptotic flatness means, (very) roughly speaking, that the metric of space-time approaches a Minkowski metric in the limit as one moves away from the source (i.e., at "infinity"). (In fact, much of the work in this game is directed toward making the previous sentence precise.) For an asymptotically flat space-time, the symmetries (and hence, the structure) of special relativity become valid in the limit at infinity. The global characterization of a system results from examining the approach of the metric to a Minkowski metric. It is through the details of this "approach" that one describes the system. (The situation has some features in

151

common with electrodynamics. The rate of approach of the electric field to zero at infinity is a measure of the total charge of the system.)

One consequence of this approach to describing a system is that the quantities one obtains (e.g., energy) will not be related, in any direct or simple way, to, say, the stress-energy of the system itself. One cannot carry the information at infinity, in any natural way, across the curvature of space-time to compare it with local information about the system being described.

One wants to define what he means by "asymptotically flat". Incorporated into such a definition will be a statement of how quickly the metric must "approach a Minkowski metric". On what rate of approach should one insist? This turns out to be a rather delicate question. We here comment on the competing factors. If the metric becomes flat too slowly, then the asymptotic behavior is not "sufficiently close to that of Minkowski space" that one recovers any global description at all. If, on the other hand, the approach to flatness is too fast, then the asymptotic metric is "so much like that of Minkowski space" that one obtains no information from the behavior of the metric. To put matters another way, a nonzero "total energy" for a system leads to a certain rate of approach of the metric to a Minkowski metric. If, in one's definition of asymptotic flatness, one requires an approach to flatness faster than this rate, then one loses the possibility of even describing systems of nonzero mass (i.e., of describing all interesting systems). Definition of asymmetric flatness represent a compromise between these two effects.

The symmetry group of flat (Minkowski) space is the Poincaré group. An analogous "asymptotic symmetry group" exists for an asymptotically flat space-time. It represents, roughly speaking, the "asymptotic motions under which the asymptotic structure is invariant". The asymptotic symmetry group for general relativity has certain features very different from the Poincaré group. There follows in intuitive discussion of why this group turns out to have the structure it does.

Let us consider again the action of Poincaré group on Minkowski space. The action of the generators of this group is represented by the Killing vectors in Minkowski space. We have seen that the general Killing vector is given by

$$\xi^a = \underline{F}^a{}_b x^b + \xi^a \tag{147}$$

where ξ^a is a constant vector field, $\underline{F}_{ab}$ is a constant antisymmetric tensor field, and x^a is the position vector relative to some origin O of space-time. what happens to (147) when we change our (arbitrary) choice of origin? Let O' be a second origin, and denote by c^a the position vector of O' relative to O. Then the position vectors of a point relative to O and O', x^a and x'^a respectively, are related by $x'^a = x^a - c^a$. Substituting, (147) takes the form $\xi^a = \underline{F}^a{}_b\, x'^b + (\xi^a + \underline{F}^a{}_b\, c^b)$.

Denote by P the Poincaré group, and let T be the sub-group consisting of translations (things generated by constant Killing vectors). Then T is nor-mal subgroup of P, and the quo-

tient group, $P/T = L$, the Lorentz group. However, P cannot be written as a product of T and L. This remark is a consequence of the discussion above. A Killing vector defines an origin-independent $\underline{F}_{ab}$, but no origin-independent ξ^a.

We now consider the asymptotic ("large" x^a) behavior of the Killing vector (147). Clearly, the first term on the right in (147) (the linear term) dominates the second (constant) term. What might we expect an "asymptotically Killing vector" to look like? Firstly, we might de-mand that it be "asymptotically linear in position", by analogy with (147). One might next be tempted to demand that "after the part asymptotically linear in position is subtracted out, what remains is asymptotically constant". But we cannot demand this! The problem is that "linear in position" is a flat-space notion – one good only in the limit in an asymptotically flat-space time. There arises an ambiguity in separating out a unique "constant part" from an "asymptotic Killing vector" because of the ambiguity in saying what "asymptotically lin-ear in position" means. Of course, as the space-time gets flatter and flatter (as one moves away from the source), this difficulty (which is associated with curvature) diminishes in importance. But the domi-nation of the constant part of (147) by the linear part increases as one moves to infinity. The consequence is that the "constant part" of an "asymptotically Killing vector" simply cannot be given any meaning. We are forced to admit as an "asymptotic Killing vector" any vector field which is asymptotically linear in position. In particular, we may add to such a vector field another which is "bounded", to obtain an-other vector field asymptotically linear in position. That is to say, the translation subgroup gets enlarged – from constant vector fields in the flat case to "bounded vector field" in the asymptotically flat case. Of course, the Lorentz part of the Poincaré group is unaffected.

Thus, the asymptotic symmetry group of general relativity is an infinite-dimensional group. There is an infinite-dimensional subgroup – called the super translations – which generalizes the finite-dimensional subgroup of translations in the Minkowski case, the quotient group is still the Lorentz group.

The remarks above are intended merely to give an intuitive feel for the structure of the asymptotic symmetry group in general relativity. In practice, the group simply pops up: it is neither introduced nor discussed in these terms.

Asymptotic structure deals with the limit at infinity. One might have thought, therefore, that the description will involve limiting procedures. Limits, of course, are inconvenient. A method has been devised for overcoming them. The reason why limits arise, of course, is that "infinity is far away". So, one "brings infinity in to a finite place, and represents it by additional points attached to space-time". How can one bring in something far away? By a conformal transformation. Thus, one introduces an unphysical metric $\tilde{g}_{ab} = \Omega^2 g_{ab}$. In order that we "shrink distances" to "bring in infinity," we choose Ω to approach zero asymptotically. With sufficient care, one can arrange matters so that the space-time manifold M, with metric admits a boundary surface of points "at infinity". Then asymptotic structure is described in terms of local structure at these points at infinity. One merely applies a conformal transformation to the relevant equations, and sees what structure emerges at the points at infinity. In fact, one goes one step further: asymptotic flatness is defined by the existence of an appropriate conformal transformation which allows a certain collection of points at infinity.

This conformal-transformation-points-at-infinity approach is convenient because it allows one to avoid complicated (and usually imprecise) limiting procedures. Put another way, one obtains the possibility of bringing local differential geometry ("local" at the points of infinity) to bear on asymptotic problems. Differential geometry replaces limits. It seems always to turn out in practice that, when something can be formulated in terms of differential geometry, it becomes simple and transparent.

In fact, there are two regimes in which one discusses asymptotic structure in general relativity: at spatial infinity and null infinity. (They refer to the mode of approach to infinity.) We shall discuss some of the special features of each regime in the following sections. One important and unsolved problem in general relativity is to relate the description of a system at spatial infinity to the description of the same system at null infinity. Essentially nothing is known. It is not even clear that asymptotic flatness in one regime implies asymptotic flatness in the other. The remarks above have been collected together because they apply to both regimes.

37. Asymptotic Structure at Spatial Infinity

We now consider the asymptotic structure of the gravitational field in the limit as one moves off to infinity "spatially". From the discussion of signal propagation in Sect. 28, one would expect that no effects of anything which happens in a compact region of space-time could have any influence on spatial infinity. Thus, the structure at spatial infinity at one "instant of time" completely determines that structure thereafter.

Consider initial data for our space-time. That is, we have a three-dimensional manifold S with positive-definite metric h_{ab} and extrinsic curvature Π^{ab}. The description of the asymptotic structure of our space-time at spatial infinity begins with the description of the asymptotic structure of this initial-data set.

Roughly speaking, we wish to require that S, h_{ab} "approach Euclidean space" asymptotically, while Π^{ab} approaches zero. To see how this statement can be formulated precisely, we must discuss some properties of Euclidean space.

Consider Euclidean space, with the usual coordinates, x, y, z, in terms of which the metric is $(dx)^2 + (dy)^3 + (dz)^2$. Set $r^2 = x^2 + y^2 + z^2)$, and introduce new coordinates

$$\bar{x} = x/r^2, \quad \bar{y} = y/r^2, \quad \bar{z} = z/r^2, \tag{148}$$

Note that the "point" $\bar{x} = \bar{y} = \bar{z} = 0$ corresponds to "infinity" ($x \to \infty$, $y \to \infty$, $z \to \infty$). Thus, approaching infinity along the x-axis (i.e., $x \to \infty$, $y = z = 0$) is, expressed in terms of barred coordinates, approaching $\bar{x} = \bar{y} = \bar{z} = 0$ along the $\bar{x}$-axis (i.e.,

$\bar{x} \to \infty$, $\bar{y} = \bar{z} = 0$). Thus, (148) suggests that we add a single point "at infinity" to Euclidean space, with $\bar{x}, \bar{y}, \bar{z}$ a coordinate system valid in a neighborhood of this point.

The next step is to rewrite the metric in terms of this barred coordinate system. First note that $d\bar{x} = (y^2 + z^2 - x^2)\,dx/r^4 - 2xy\,dy/r^4 - 2xz\,dz/r^4$, etc. for $d\bar{y}$, $d\bar{z}$. Hence,

$$r^4(\,d\bar{x}^2 + d\bar{y}^2 + d\bar{z}^) = (\,dx^2 + dy^2 + dz^2) \qquad (149)$$

Thus, if we wish to regard $\bar{x}$, $\bar{y}$, $\bar{z}$ as "good" coordinates near infinity, then the original metric of our Euclidean space becomes badly behaved near "infinity", $\bar{x} = \bar{y} = \bar{z} = 0$. This, of course, is to be expected. After all, infinity is a long way away (according to the original Euclidean metric). If we insist that there be a point at infinity, then, in order that this point be infinitely far away from other points of our space, it is necessary that the Euclidean metric grow without bound near this point at infinity.

Denote by h_{ab} the original Euclidean metric. Then, from (149), the metric $\tilde{h}_{ab} = \Omega^2 h_{ab}$ is well-behaved near infinity (in the coordinates valid in that region: $\bar{x}$, $\bar{y}$, $\bar{z}$), provided we choose $\Omega = 1/r^2$. This, too, is what we might have expected. In order to obtain a metric which is well-behaved near infinity, we must perform a conformal transformation using a conformal factor Ω which approaches zero at infinity. In this way, we "shrink distances", bringing infinity in to a finite distance from other points of our space.

How shall we express, in a coordinate-independent way, the asymptotic behavior of the conformal factor, $\Omega = 1/r^2$? Set $\bar{r}^2 = \bar{x}^2 + \bar{y}^2 + \bar{z}^2$. Then, from (148), $\Omega = \bar{r}^2$. Next, note that $\bar{r}$ represents, geometrically, the distance from infinity, as measured by the metric $\tilde{h}_{ab}$. Hence, at the point at infinity, $\Omega = 0$, $\tilde{D}_a\Omega = 0$, and $\tilde{D}_a\tilde{D}_b\Omega = 2\tilde{h}_{ab}$, where $\tilde{D}_a$ is the derivative (with respect to $\tilde{h}_{ab}$). it is clear that these conditions describe the asymptotic behavior of the conformal factor.

The discussion above can be regarded as motivation for the simple and precise definition which follows. We have deliberately been sloppy in our discussion of Euclidean space, for the asymptotic structure of Euclidean space, suggested above, is expressed very neatly by our definition. The idea is to define asymptotic flatness of a space by the existence of a "conformal completion" which has the essential properties of the conformal completion described above for Euclidean space.

Let S be a three-dimensional manifold with positive-definite metric h_{ab}. This S, h_{ab}, will be said to be *asymptotically flat* if there exists a manifold $\tilde{S} = S \cup \Lambda$, where Λ is an abstract point (i.e., S is a sub manifold of $\widetilde{S}$, with $\widetilde{S} - S$ a single point), with metric $\tilde{h}_{ab}$ (on $\widetilde{S}$) such that

1. On S, $\tilde{h}_{ab} = \Omega^2 h_{ab}$, where Ω is a scalar field on S.

2. At the point Λ, $\Omega = 0$, $\widetilde{D}_a \Omega = 0$, and $\widetilde{D}_a \widetilde{D}_b \Omega = 2\tilde{h}_{ab}$.

(Note: In our discussion of Euclidean space, the conformal factor, $1/r^2$, became singular at the origin. Clearly, the same discussion could have been carried out – less conveniently – using a conformal factor which is well-behaved at the origin.)

We now claim that Euclidean 3-space is asymptotically flat in the sense above. It should be clear that asymptotic flatness is a reasonable replacement for the intuitive statement that the metric "becomes Euclidean sufficiently quickly at infinity".

We now return to our initial data set, S, h_{ab}, Π^{ab}. We want to define asymptotic flatness for such a thing. We first require that S, h_{ab}, be asymptotically flat in the sense above. Intuitively, this represents a condition on the "geometry". What remains is to impose condition on the "fields", including, in particular, "Π^{ab}". It turns out that the appropriate conditions are that the tensor fields $\widetilde{\Pi}^{ab}$, $\Omega^{1/2}\widetilde{R}_{ab}$, and $\Omega^{-1/2}(\widetilde{D}_a \widetilde{D}_b \Omega - 2\tilde{h}_{ab})$ have limits at the point Λ, where we have assigned to Π^{ab} dimensions $\sec^{-11}$ (i.e., $\widetilde{\Pi}^{ab} = \Omega^{-3}\Pi^{ab}$). Thus, in particular, we require that Π^{ab} approach zero at infinity, at a certain rate. The values of these three tensors at the point Λ at infinity represent the asymptotic gravitational field. The asymptotic quantities characteristic of our system as a whole are defined in terms of these three tensors.

This, then, is the framework of the discussion of asymptotic structure at spatial infinity. One chooses an initial-data surface which is asymptotically flat, introduces a point Λ at infinity, and obtain certain tensors at that point which describe the asymptotic behavior of the gravitational field.

One is interested, of course, in the asymptotic structure of spacetime, not in the asymptotic structure of "space". In other words, one would want the description above to be independent of the choice of initial surface – or, at least, to have a simple dependence on the choice of surface. We next discuss how one investigates dependence on the choice of surface.

The equations which govern the evolution of initial data are (136) and (137), where φ, the evolution function, is at our disposal. It describes the distance one moves (normally) off the initial surface S at each point of S. Suppose our initial S, h_{ab}, Π^{ab}, were asymptotically flat. It is clear, at least intuitively, that for certain choices of φ, the next surface will not be asymptotically flat (e.g., choose φ to become infinite, quickly, near infinity). Thus, in order that asymptotic flatness (of S, h_{ab}, Π^{ab}) be preserved during the evolution, we must be selective in our choice of φ. What conditions must we impose on φ in order that asymptotic flatness be preserved? To find out, we return to the case of Minkowski space.

Suppose that our initial S were a three-dimensional plane in Minkowski space, i.e., suppose h_{ab} were flat and Π^{ab} vanished. For what choices of φ are the succeeding surfaces also planes? The answer is immediate from (136) and (137): we must have $D^a D^b \varphi = 0$. In other words, φ must have the form

$$\varphi = s_a x^a + s \qquad (150)$$

where s_a is a constant vector field, s is a constant scalar field, and x^a is the position vector relative to some origin (in our Euclidean space). Geometrically, $\varphi = s = $ constant, corresponds to an infinitesimal time-translation of our surface, while $\varphi = s_a x^a$ corresponds to an infinitesimal boost. Thus, in the general curved case, we must demand that, asymptotically, φ behave like (150). We must characterize this asymptotic behavior locally, in term of behavior at Λ. Let φ have dimensions sec, so $\widetilde{\varphi} = \Omega \varphi$. Then (150) gives $\widetilde{\varphi} = r^{-2} s_a x^a + r^{-2} s = s_a \widetilde{x}^a + \widetilde{r}^2 s$. Hence, at the point Λ,

$$\widetilde{\varphi} = 0 \quad \widetilde{D}_a \widetilde{\varphi} = s_a \quad \widetilde{D}_a \widetilde{D}_b \widetilde{\varphi} = 2\, s\, \widetilde{h}_{ab} \qquad (151)$$

In words, $\widetilde{\varphi}$ vanishes at Λ, while the derivative of $\widetilde{\varphi}$ at Λ gives the amount of boost represented by φ, and the second derivative of $\widetilde{\varphi}$ at Λ is a multiple of the metric, the multiplicative factor giving the amount of time-translation represented by φ.

Thus, to preserve asymptotic flatness in the curved case, one would be tempted to insist that φ be such that, at Λ, the conditions (151) hold. It turns out that, whereas the first two conditions (151) can indeed be imposed in the presence of curvature, the third cannot. This remark is suggested by the following equation

$$\widetilde{D}_{[a} \widetilde{D}_{b]} \widetilde{D}_c \widetilde{\varphi} = \tfrac{1}{2} \widetilde{\mathscr{R}}_{abc}{}^d \widetilde{D}_d \widetilde{\varphi} = \tfrac{1}{2} \widetilde{\mathscr{R}}_{abc}{}^d s_d \qquad (152)$$

where, in the second step, we have used the second equation (151). It is perhaps not surprising, from (152), that, when $\widetilde{\mathscr{R}}_{abcd} = 0$ and $s_a = 0$, we are not free to choose $\widetilde{D}_a \widetilde{D}_b \widetilde{\varphi}$, at Λ, to be a multiple of $\widetilde{h}_{ab}$. In words, in the presence of curvature and a boost, time-translations can no longer be cleanly and unambiguously distinguished. What one does, in light of (152) is to enlarge the time-translations to encompass possibilities more general than $D_a D_b = 2\,s\,h_{ab}$. Here is the "enlargement of the translation subgroup" which appears in the asymptotic symmetry group – the phenomenon discussed in Sect. 36.

We have now discussed boosts (which are represented by vectors at Λ, just as in the flat case, and which, therefore, play the same role as in flat space), time-translations (which, because of (152), must be enlarged to an infinite-dimensional group). Spatial rotations, of course, are merely represented by rotations within the tangent space of Λ (The reason: Perform a rotation on x, y, z in Euclidean space, and see what it does to $\bar{x}, \bar{y}, \bar{z}$ and hence to the tangent space at Λ.) Where do the spatial translations show up in the formalism?

To answer this question, we return to the Euclidean case. Recall that we choose for our conformal factor $\Omega = 1/r^2$, where r is the distance from origin. A

translation (spatial) can be represented by a change in this choice of origin and, hence, by a change in choice of conformal factor. Let O' be a second origin, with position vector c^a with respect to O. Then, writing r' for the distance from O' we have $(r')^2 = (x^a - c^a)(x_a - c_a) = r^2 - 2c_a x^a + c^a c_a$. Hence, the conformal factor which results from choosing O' as our origin is $\Omega' = (r^2 - 2c_a x^a + c^a c_a)^{-1}$. Set $\Omega = \Omega'\omega$, so $\omega = 1 - 2c_a x^a r^{-2} = 1 - 2c_a \bar{x}^a + (c_a c^a)\bar{r}^2$ That is to say, the transition from conformal factor Ω to conformal factor Ω' is is accomplished by a conformal transformation with conformal factor $\omega = 1 - 2c_a \bar{x}^a + c_a c^a \bar{r}^2$. This ω is one at Λ, while its derivative at Λ represents the spatial translation (i.e., c^a). Similarly, the behavior of fields at Λ in the general (curved) case under further conformal transformations represents the effects of spatial translations.

The remarks above are intended as a survey of what is, in detail, a rather technical subject. We summarize. Asymptotic structure at spatial infinity is described in terms of an initial-data set for the space-time. We attach to this three-dimensional manifold a single point Λ at infinity, and extend the metric to this point by means of a conformal transformation. Asymptotic flatness of our initial data set is defined by the existence of such a conformal completion, such that certain tensor fields assume finite values at Λ. These tensors at Λ describe the asymptotic gravitational field. Asymptotic boosts and time-translations show up in the asymptotic behavior of the evolution function φ. Asymptotic rotations show up as rotations of the tangent space of Λ. Asymptotic spatial translation show up as conformal transformations. All these motions, taken together, constitute the asymptotic symmetry group at spatial infinity.

38. Asymptotic Structure at Null Infinity

The second regime for the study of asymptotic structure is in the limit at null infinity. We briefly summarize the set-up.

Consider Minkowski space in spherical coordinates: $-dt^2 + dr^2 + r^2(d\theta^2 + \sin^2\theta\,d\varphi^2)$. Replace the coordinates r, t by hew coordinates $u = t + r$ and $v = t - r$, so the metric becomes $-du\,dv + (1/4)(u - v)^2(d\theta^2 + \sin^2\theta\,d\varphi^2)$. Geometrically, the surfaces of constant u are past null cones, centered on the t-axis, while the surfaces $v = $ const. are future-directed null cones. The two limits with which we shall be concerned are future null infinity 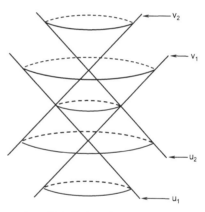 ($u \to \infty$, with v, θ, φ finite) and past null infinity ($v \to -\infty$, with u, θ, and φ finite)

The idea is to do here for Minkowski space what we did, in Sect. 37, for Euclidean space. We cleverly select a system of coordinates in terms of which there is a natural conformal transformation on our metric and a natural "boundary at infinity". We then describe the situation in coordinate-independent terms. Finally, we use the existence of appropriate conformal completion as the definition of asymptotic flatness.

We must introduce still another coordinate transformation in Minkowski space such that our asymptotic limits ($u \to \infty$, with v, θ, φ finite; $v \to -\infty$, with u, θ, φ finite) appear at finite coordinate values. Set $u = \tan p$, $v = \tan q$. Then $u \to \infty$ corresponds to $p = \pi/2$, and $v \to$ to . In terms of these coordinates, our metric takes the form

161

$$\cos^{-2} p \cos^{-2} q \left[- \mathrm{d}p \, \mathrm{d}q + \tfrac{1}{4} \sin^2(p - q)(\mathrm{d}\theta^2 + \sin^2 \theta \, \mathrm{d}\varphi^2) \right]$$

Geometrically, we have a null, three-dimensional surface $\mathscr{J}^+(p = \pi/2)$, and a second null, three-dimensional surface $\mathscr{J}^-(q = -\pi/2)$ attached to our space-time. These are the surfaces at "null infinity." Asymptotic structure at null infinity is to be describes as local structure on these surfaces. Our metric, (113) is not, of course, well-behaved on

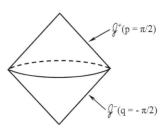

$\mathscr{J}^+$ and $\mathscr{J}^-$. However, $\Omega^2 g_{ab}$ is well-behaved, where $\Omega = \cos p \cos q$. Note that the conformal factor Ω vanishes at infinity (as it must, in order to bring in infinity to a finite surface). Finally, note that, on $\mathscr{J}^\pm$, $\widetilde{\nabla}_a \Omega$, is null and nonzero.

These observations about Minkowski space having been made, we are prepared to define asymptotic flatness of a space-time at null infinity. Let M, g_{ab} be a space-time. This space-time is said to be asymptotically flat (at null infinity) if there exists a manifold $\widetilde{M} = M \cup \mathscr{J}^+ \cup \mathscr{J}^-$, where $\mathscr{J}^+$ and $\mathscr{J}^-$ are boundary surfaces (more formally, we should introduce a precise notion here: that of manifold-with-boundary), which metric such that

1. On M, $\widetilde{g}_{ab} = \Omega^2 g_{ab}$, where Ω is some scalar field on $\widetilde{M}$.

2. $\mathscr{J}^\pm$ are null surfaces.

3. On $\mathscr{J}^\pm$, $\Omega = 0$, and $\widetilde{\nabla}_a \Omega$ is a nonzero null vector.

We now assert, from our earlier remarks, that Minkowski space is asymptotically flat in this sense.

How does the description of asymptotic structure at null infinity compare with that at spatial infinity? We remarked earlier that nothing ever happens at spatial infinity. Quite the reverse is the case at null infinity. From our remarks on signal propagation, it is clear that radiation escapes to null infinity: it is registered on $\mathscr{J}^+$. Thus, null infinity (the "radiation zone") is the natural asymptotic limit for the discussion of radiation. On the other hand, conservation laws are more easily discussed at spatial infinity.

There are numerous unsolved problems in general relativity concerning asymptotic structure. Perhaps the most interesting are those concerning the relation between the two asymptotic descriptions. Essentially all questions of the type "If such-and-such is true at special infinity, does it follow that such-and-such is true at null infinity?" are unanswered. It is not even known whether not asymptotic flatness in one of the two senses implies asymptotic flatness in the other!

39. Linearization

To what does Einstein's equation reduce when the metric of space-time is "nearly flat". More precisely, suppose we set $g_{ab} = \eta_{ab} + \gamma_{ab}$, where η_{ab} is a Minkowski (flat) metric, and γ_{ab} is "small". We could then imagine expanding Einstein's equation to first order in the perturbation, γ_{ab}. The resulting equations would provide an approximate description of "weak gravitational fields" (i.e., fields sufficiently weak that they do not require that the geometry depart significantly from flatness). These equations provide a simple picture of the behavior of gravitational fields in this limit. They are called the linearized Einstein equation. we obtain the linearized equation in the present section.

It is convenient to adopt an approach in which one does not have to keep track of the orders of various terms in the perturbation expansion. Let $g_{ab}(\lambda)$ be a one-parameter family of metrics on a fixed manifold M, where the parameter λ has range, say $[0, 1)$. Let $g_{ab}(0)$ be a flat metric (i.e., with vanishing Riemann tensor). The plan is to evaluate $\mathrm{d}/\mathrm{d}\lambda$ of various quantities at $\lambda = 0$. Thus, by introducing this "expansion parameter λ," and taking derivatives with respect to λ, we automatically keep the appropriate terms in our perturbation expansion.

For each value of λ, denote by $\overset{\lambda}{\nabla}_a$ the derivative operator defined by the metric $g_{ab}(\lambda)$. Fix a (λ–independent) vector field k_b on M. Then

$$\frac{\mathrm{d}}{\mathrm{d}\lambda}\left(\overset{\lambda}{\nabla}_a k_b\right)\Bigg|_{\lambda=0} = -C^m{}_{ab}\, k_m \tag{153}$$

for some tensor field $C^m{}_{ab} = C^m{}_{(ab)}$. (Proof: Note that the left side of (153) is linear in k_b, and that the anti symmetric part of the left side vanishes.) Our first task is to evaluate this $C^m{}_{ab}$. Setting $g_{ab} = g_{ab}(0)$, $\nabla_a = \overset{0}{\nabla}_a$, we have

$$0 = \frac{\mathrm{d}}{\mathrm{d}\lambda}\left[\overset{\lambda}{\nabla}_a g_{bc}(\lambda)\right]\Bigg|_{\lambda=0}$$

$$= -C^m{}_{ab}\, g_{mc} - C^m{}_{ac}\, g_{bm} + \nabla_a\left[\frac{\mathrm{d}}{\mathrm{d}\lambda}g_{bc}(\lambda)\,\Big|_{\lambda=0}\right] \qquad (154)$$

where, in the first step, we have used the fact that $\overset{\lambda}{\nabla}_a g_{bc}(\lambda)$ vanishes for all λ, and, in the second step, we have used (153). Set $\gamma_{ab} = (\mathrm{d}/\,\mathrm{d}\lambda)g_{ab}(\lambda)|_{\lambda=0}$. Then the solution of (154) in

$$C^m{}_{ab} = \tfrac{1}{2}g^{mn}(\nabla_a\gamma_{bn} + \nabla_b\gamma_{an} - \nabla_n\gamma_{ab}) \qquad (155)$$

Think of $C^m{}_{ab}$ as representing the "λ–rate of change of the derivative operator, evaluated at $\lambda = 0$". This $C^m{}_{ab}$ is given by (155).

For each λ, we have a Riemann tensor, $R_{abc}{}^d(\lambda)$. Of course, $R_{abc}{}^d(0)$ vanishes, for we choose our metric to be flat at $\lambda = 0$. However, the linearized Riemann tensor, $(\mathrm{d}/\,\mathrm{d}\lambda)R_{abc}{}^d(\lambda)|_{\lambda=0}$, will not vanish in general. This quantity represents the curvature which results, to first order, from our perturbation of the metric away from a flat metric. To evaluate it, we proceed as follows. Let k_c be a fixed vector field on M. Then, for each value of λ, $\overset{\lambda}{\nabla}_{[a}\overset{\lambda}{\nabla}_{b]}k_c = \tfrac{1}{2}R_{abc}{}^d(\lambda)\,k_d$. Taking $\mathrm{d}/\,\mathrm{d}\lambda$, and evaluating at $\lambda = 0$,

$$\frac{1}{2}\frac{\mathrm{d}}{\mathrm{d}\lambda}R_{abc}{}^d(\lambda)\,\Big|_{\lambda=0}\,k_d = \left[\frac{\mathrm{d}}{\mathrm{d}\lambda}\overset{\lambda}{\nabla}_{[a}\overset{\lambda}{\nabla}_{b]}k_c\right]_{\lambda=0}$$

$$= -C^d{}_{[ab]}\overset{\lambda}{\nabla}_d k_c\,\Big|_{\lambda=0} - C^d{}_{c[a}\overset{\lambda}{\nabla}_{b]}k_d\,\Big|_{\lambda=0} + \nabla_{[a}\left(\frac{\mathrm{d}}{\mathrm{d}\lambda}\overset{\lambda}{\nabla}_{b]}k_c\,\Big|_{\lambda=0}\right)$$

$$= -C^d{}_{[ab]}\nabla_d k_c - C^d{}_{c[a}\nabla_{b]}k_d + \nabla_{[a}(-C^d{}_{b]c}k_d)$$

$$= -C^d{}_{c[a}\nabla_{b]}k_d - \nabla_{[c}C^d{}_{b]c}k_d - C^d{}_{c[b}\nabla_{a]}k_d$$

$$= -k_d\nabla_{[a}C^d{}_{b]c}$$

where, in the second step, we have used (153), in the third step, we have again used (153), in the fourth step, we have expanded and used $C^d{}_{[ab]} = 0$ and, in the fifth step, we have canceled the first and last terms. Since k_d is arbitrary,

$$\frac{\mathrm{d}}{\mathrm{d}\lambda}R_{abc}{}^d(\lambda)\,\Big|_{\lambda=0} = -2\nabla_{[a}C^d{}_{b]c} \qquad (156)$$

Now substitute (155). (We shall hereafter raise and lower indices with g_{ab}, the "unperturbed metric".)

$$\frac{\mathrm{d}}{\mathrm{d}\lambda}R_{abc}{}^{d}(\lambda)\,\big|_{\lambda=0}=\nabla^{d}\nabla_{[a}\gamma_{b]c}-\nabla_{c}\nabla_{[a}\gamma_{b]}{}^{d}\qquad(157)$$

This is the expression, to first order, for the curvature which results from a perturbation of the metric of flat space. Contracting (157) once, (with $\gamma=\gamma^{m}{}_{m}$)

$$\frac{\mathrm{d}}{\mathrm{d}\lambda}R_{ab}(\lambda)\,\big|_{\lambda=0}=\nabla_{(a}\nabla^{m}\gamma_{b)m}-\frac{1}{2}\nabla^{m}\nabla_{m}\gamma_{ab}-\frac{1}{2}\nabla_{a}\nabla_{b}\gamma\qquad(158)$$

Now we suppose our family of metrics, $g_{ab}(\lambda)$, represents a family of solutions of Einstein's equation, so we have a stress-energy, $T_{ab}(\lambda)$. Of course, $T_{ab}(\lambda)=0$, since our initial metric is flat. The quantity $(\mathrm{d}/\mathrm{d}\lambda)T_{ab}(\lambda)|_{\lambda=0}$ thus represents the "first-order stress-energy of matter". From (158), Einstein's equation becomes

$$\begin{aligned}8\pi G\frac{\mathrm{d}}{\mathrm{d}\lambda}T_{ab}(\lambda)\,\Big|_{\lambda=0}&=\nabla_{(a}\nabla^{m}\gamma_{b)m}-\frac{1}{2}\nabla^{m}\nabla_{m}\gamma_{ab}-\frac{1}{2}\nabla_{a}\nabla_{b}\gamma\\&\quad-\frac{1}{2}g_{ab}\nabla^{m}\nabla^{n}\gamma_{mn}+\frac{1}{2}g_{ab}\nabla^{m}\nabla_{m}\gamma\end{aligned}\qquad(159)$$

Eqn, (159) can be simplified somewhat. The steps are very closely analogous to corresponding steps in the discussion of potentials for electromagnetism, so we begin with that case. Recall equations: $\nabla^{[a}F^{bc]}=0$, $\nabla_{b}F^{ab}=J^{a}$. We restrict consideration to flat space. The first equation above implies that

$$F^{ab}=\nabla^{[a}A^{b]}\qquad(160)$$

for some vector field A^{a} (called the *vector potential*) in our flat space. Substituting, the second Maxwell equation gives

$$-\frac{1}{2}\nabla^{m}\nabla_{m}A^{a}+\frac{1}{2}\nabla^{a}(\nabla_{m}A^{m})=J^{a}\qquad(161)$$

Eqn. (160) does not determine this potential A^{a} uniquely from F^{ab}. In fact, A^{a} and $A'^{a}=A^{a}+\nabla^{a}\psi$, where ψ is any scalar field, which defines, via (160), the same F^{ab}. This $A'^{a}=A^{a}+\nabla^{a}\psi$ is called a *gauge transformation* on the vector potential. We have

$$\nabla_{a}A'^{a}=\nabla_{a}A^{a}+\nabla_{a}\nabla^{a}\psi\qquad(162)$$

Setting the left side of (162) equal to zero, we obtain a wave equation with source for ψ. This equation has a solution. Thus, by means of a gauge transformation, we can obtain a vector potential A^{a} satisfying (160) and also

$$\nabla_a A^a = 0 \tag{163}$$

Eqn. (163) is called the *Lorentz gauge condition*. Thus, we utilize the freedom available in gauge transformations to obtain a vector potential satisfying the subsidiary condition (163). The advantage is that (161) now simplifies to

$$-\frac{1}{2}\nabla^m \nabla_m A^a = J^a \tag{164}$$

The electromagnetic situation above is completely analogous to the gravitational situation. We regard the vector potential A^a as analogous to the "gravitational potential", γ_{ab}. The electromagnetic field, F^{ab}, is analogous to the left side of (157) – the linearized Riemann tensor. Then (160) is analogous to (157). The electromagnetic charge-current, J^a, is analogous to the linearized stress-energy, $(\mathrm{d}/\mathrm{d}\lambda)T_{ab}(\lambda)|_{\lambda=0}$. (That is, the source for the electromagnetic field is analogous to the source for the gravitational field.) Thus, (159) is analogous to (161).

What in the gravitational case is analogous to the electromagnetic gauge transformations and Lorentz gauge? First note that, replacing γ_{ab} on the right in (157) by $\gamma'_{ab} = \gamma_{ab} + \nabla_{(a}\xi_{b)}$, where ξ_b is any vector field, the right side of (157) is unchanged. Thus, we regard $\gamma'_{ab} = \gamma_{ab} + \nabla_{(a}\xi_{b)}$ as representing "gravitational gauge transformations", with ξ_b analogous to ψ. These gravitational gauge transformations change the linearized metric (γ_{ab}) without changing the linearized curvature. Next, note that

$$\nabla_b(\gamma'^{ab} - \tfrac{1}{2}\gamma' g^{ab}) = \nabla_b(\gamma^{ab} - \tfrac{1}{2}\gamma g^{ab}) + \tfrac{1}{2}\nabla^m \nabla_m \xi^a \tag{165}$$

Thus, we can always perform a gauge transformation such that

$$\nabla_b(\gamma^{ab} - \tfrac{1}{2}\gamma g^{ab}) = 0 \tag{166}$$

Eqn. (159) then becomes

$$8\pi G \frac{\mathrm{d}}{\mathrm{d}\lambda}T_{ab}(\lambda)\,\Big|_{\lambda=0} = -\frac{1}{2}\nabla^m \nabla_m(\gamma_{ab} - \frac{1}{2}\gamma g_{ab}) \tag{167}$$

We regard (165), (166), and (167) as analogous to (162), (163), and (164), respectively. Note that (167) is essentially a wave equation, with the linearized metric.

We summarize with the table below:

The field is	$\frac{\mathrm{d}}{\mathrm{d}\lambda}R_{abcd}\vert_{\lambda=0}$	F_{ab}
We introduce a potential	γ_{ab}	A_a
in terms of which the field is	$\nabla_d\nabla_{[a}\gamma_{b]c} - \nabla_c\nabla_{[a}\gamma_{b]d}$	$\nabla_{[a}A_{b]}$
The potential is determined by the field only up to	$\gamma_{ab} \to \gamma_{ab} + \nabla_{(a}\xi_{b)}$	$A_a \to A_a + \nabla_a\psi$
For arbitrary	ξ_b	ψ
These are the gauge transformations. One can by means of appropriate gauge transformations choose a potential satisfying	$\nabla_b(\gamma^{ab} - \frac{1}{2}\gamma g^{ab}) = 0$	$\nabla_a A^a = 0$
The source for the field is	$\frac{\mathrm{d}}{\mathrm{d}\lambda}T_{ab}(\lambda)\vert_{\lambda=0}$	J^a

The source "generates potential" via	$8\pi G\frac{\mathrm{d}}{\mathrm{d}\lambda}T_{ab}(\lambda)\Big\vert_{\lambda=0} =$ $\nabla_{(a}\nabla^m\gamma_{b)m} -$ $\frac{1}{2}\nabla^m\nabla_m\gamma_{ab} -$ $\frac{1}{2}\nabla_a\nabla_b\gamma -$ $\frac{1}{2}g_{ab}\nabla^m\nabla^n\gamma_{mn} +$ $\frac{1}{2}g_{ab}\nabla^m\nabla_m\gamma$	$J_a =$ $-\frac{1}{2}\nabla^m\nabla_m A_a +$ $\frac{1}{2}\nabla_a(\nabla^m A_m)$
If, however, the gauge condition is satisfied, this simplifies to	$8\pi G\frac{\mathrm{d}}{\mathrm{d}\lambda}T_{ab}(\lambda)\Big\vert_{\lambda=0} =$ $-\frac{1}{2}\nabla^m\nabla_m(\gamma_{ab} - \frac{1}{2}\gamma g_{ab})$	$J_a =$ $-\frac{1}{2}\nabla^m\nabla_m A_a$

The advantage of introducing potentials is in both cases the same. One in this way obtains equations of the form $\nabla^m\nabla_m$ (potentials) = sources. Such equations are easy to solve, e.g., by Green's functions.

40. Quantization of Gravitational Field: Introduction

We have been discussing the general theory of relativity. Perhaps the three fundamental principles of this theory are the following; i) The events of space-time (all events, past, present, and future) are assembled into a four-dimensional manifold. The description of physics is in terms in of fields on this manifold. ii) There is a metric tensor field g_{ab}, on this manifold. The metric simultaneously describes the geometry (results of space and time measurements) of space-time and the effect of gravitation. iii) Matter in space-time produces a certain tensor field which causes the metric of space-time to exhibit curvature. One of the central features of the general theory of relativity (a feature we have, perhaps, not stressed strongly enough) is that the theory claims to incorporate within its structure all of physics. Where there's "physics", there's stress-energy, and hence there's curvature of space-time. The full apparatus of general relativity must, at least in principle (though almost never in practice!) be brought into play in the discussion of any physical phenomenon. General relativity claims a universality over other areas of physics. (Note the word "claims". The entire theory could, of course, differ substantially from the way Nature chooses to behave.)

There exists at least one other theory of physics with a similar claim to universality: quantum theory. In my opinion, the fundamental principles of quantum theory are: i) The states of a system are described in terms of a Hilbert space (more specifically, by rays in a Hilbert space), and ii) the attributes (properties of, measurements on, etc.) of the system are described in terms operators on that Hilbert space. Perhaps quantum theory can be viewed as the insistence that every theory of physics be formulated according to the principles above.

That there is a problem here should now be clear. The general theory of relativity is not formulated in the terms demanded by quantum theory. One seeks, therefore, a modification of the theory to obtain consistency with the principles of quantum theory: one seeks to quantized the general theory of relativity. The problem, then, is to write down a theory which is both "quantum-theory-looking" and "general-relativity-looking". This problem, to which a great deal of effort and many clever ideas have been directed, remains unsolved. We shall, in the next few sections, discuss a few of the approaches to this problem. It should be emphasized that whether or not a collection of sentences and equations represents a "solution" to this problem is, for the most part, an aesthetic question.

What features might one expect to appear in a quantum theory of gravitation? It is common that concepts in a classical theory which are "sharp" become "fuzzed out" on quantization. For example, for a particle approaching a potential barrier, classically, the particle either reflects or is transmitted, while, in quantum theory, there is merely a distribution in probabilities for various outcomes. The primary candidate for something to be fuzzed out on quantization of general relativity is the point events of space-time. One might expect that these events will lost their significance – i.e., will reappear only in the classical limit of the quantum theory of gravitation. This expectation is suggested, for example, by the following remark. Suppose we build a probe of some sort which makes measurements in a very small region of space-time (or, in the limit, at a single event of space-time). Then our probe must be at least as small as the region over which it makes measurements. But the uncertainty principle in quantum theory suggests that very small instruments must contain particles of high momentum – hence, high energy. But, if our probe is to have a large stress-energy, then by Einstein's equation, it must be responsible for large curvatures of space-time. In other words, a significant distortion of space-time in the region being measured will result from introducing our probe. Thus, it appears that point events will lose their operational significance under quantization. But it usually happens in physics that, when a concept looses operational significance, that loss is reflected in the mathematical formulation of the theory.

A second concept from general relativity one might expect to be "fuzzed out" by quantization is the metric. There will not, presumably, be one specific metric of space-time, but some probability distribution of possible metrics. This "smearing out of the metric" might be expected to have significant physical consequences. For example, the divergences which arise in quantum field theory come about, at least in part, because integrals in momentum space extend to arbitrary large momenta. One uses "cutoffs" in momentum to obtain finite results.

If the metric were "smeared out", one might expect this to result in natural cutoffs on such integrals. One might expect the divergence difficulties associated with quantum field theories to, at least, become less severe in the presence of a quantized metric. Furthermore, the singularities we have seen in general relativity might also be expected to disappear. The smoothing out from quantum theory could result in a smoothing over of these singularities. (Analogous phenomenon in atomic physics: Classically, an electron orbiting a point nucleus radiates, spirals inward, and eventually hits the nucleus. In quantum theory, this singularity disappears.)

The approaches to quantization of general relativity are normally based on analogies with quantum theories we understand: quantum electrodynamics, Schrödinger quantum mechanics for a particle, etc.

41. Linearized Approach to Quantization

Consider the linearized Einstein equations, (166) and (167). We have seen in Sect. 39 that this approach to general relativity results in a set of equations on a field γ_{ab} in flat space which bears a very close resemblance to Maxwell's equations of electrodynamics. The vector potential A_a for the electromagnetic field is replaced by the "potential for the gravitational field", γ_{ab}. These are both tensor fields in flat space-time. In some sense, the linearized Einstein equations represent an approximation to the full equations of general relativity.

The approach to quantization to be discusses in this section is based on the following idea. One regards γ_{ab} as just another classical field (on the same footing with, say, the electromagnetic vector potential). One attempts to use the conventional techniques of quantum field theory on this γ_{ab}. That is to say, one extends the analogy between electrodynamics and linearized general relativity to a quantization program for the latter. Using quantum electrodynamics as a model, one attempts to construct a "quantum gravidynamics".

In this section, we shall first summarize, in very broad and vague terms, the setting of quantum electrodynamics. We then remark that similar techniques could be applied to the linearized Einstein equation. Finally, we make some general comments on the resulting "quantum theory of gravitation".

There are two stages leading to quantum electrodynamics. In first, one obtains the theory for free photons (the quantized version of the classical theory described by $\nabla^m \nabla_m A^a = 0$, $\nabla_a A^a = 0$). Next, one introduces interactions. For the free case, one proceeds, roughly, as follows. Consider the real (infinite-dimensional) vector space of (asymptotically well-behaved) solutions of Maxwell's equations with $J^a = 0$. One introduces on this vector space a suitable norm and a suitable complex structure. It thus becomes a Hilbert space H. This H represents the Hilbert space of one-photon states of (source-free) Maxwell field. Next, one extends this description to states with many photons.

$$F = H^0 + H^1 + H^2 + H^3 + \cdots \qquad (168)$$

where the superscripts denote "powers" of H. (More precisely, H^0 is the complexes, $H^1 = H$, H^2 is the tensor products of H with itself, H^3 the tensor product of H with H^2, etc. The sums are direct sums of Hilbert spaces.) This F, the Fock space, represents the states of the system (without sources). An element of H^n represents a state with n photons (and an element of H^0 a vacuum (zero photon) state). Thus, the general element of F consists of a linear combination of states with various numbers of photons. There are defined operators on F representing such things as "numbers of photons", "energy-momentum", etc. This, the theory of free (non-interacting) photons, is not very interesting, because nothing much happens.

One now introduces interactions. These are described by certain operators (on F, the Fock space for electrodynamics, and also on the Fock spaces for the other particles of interest, e.g., electrons). These interaction operators allow for the possibility of translations in which numbers of photons change (while, of course, numbers of other types of particles can also change). Thus, with the introduction of an interaction, one has the possibility of particle reactions' taking place. In this way, e.g., electron-photon scattering cross sections can be calculated, and comparison made with experiment.

Essentially the same program goes through, with little change, for the linearized Einstein equation. One introduces the Fock space of free gravitation states. For the interaction, the gravitons couple to the stress-energy of particles rather than (in the electromagnetic case) the charge. One calculates scattering processes, etc. involving gravitons. The proposal, then, is that one regard the result as representing a "quantization of general relativity".

There is certainty a sense in which the program above departs from the spirit of general relativity. One could, of course, criticize it on the grounds that it deals only with the linearized equations – not the full Einstein equation. This, however, is a deficiency only of our brief description – not of the program itself. One could just as well consider also the higher order terms in the perturbation expansion (i.e., in Sect. 39, one could take $\mathrm{d}^2/\mathrm{d}\lambda^2$, $\mathrm{d}^3/\mathrm{d}\lambda^3$, etc., at $\lambda = 0$, of Einstein's equation). These corrections would represent further possible interactions – they would be gravitation-gravitation interactions. Thus, one would regard the nonlinearity of Einstein's equation as allowing for the possibility of "gravitational field produced by gravitational field itself". Quantum-mechanically, gravitons create gravitons. The more terms included as interactions from the perturbation expansion, presumably, the closer the resulting field theory would approximate general relativity.

In my view, more serious objections are possible. A physical theory consists, of course, of more than merely the equations of that theory. In particular, general relativity consists of more than Einstein's equation. There is in addition to the equations, an overlay of concepts, attitudes, prejudices, etc. The concepts play at least as great a role in what the theory "is" as the equations. In general relativity, for example, there is the notion of assembling all possible events into a manifold. There is the notion of the metric on this manifold – an object with direct physical significance as giving the result of space and time measurements, and more indirect physical significance concerning gravitation. In short, general relativity is an integral part of what might be called the "space-time view of physics".

Where are these concepts from general relativity in the linearized version of quantized gravitation? One sees, at least, the rudiments of Einstein's equation, but, in my opinion, not the sense of the general theory of relativity. This is not to say, of course, that the linearized program is wrong. What it does seem to imply is that, if Nature behaves as described by this approach, then the general theory of relativity has been an unfortunate – and expensive in terms of time and effort – detour.

42. Canonical Approach to Quantization

As an alternative to the linearized approach, we now discuss the canonical approach to quantization. Perhaps this approach displays a greater respect for the integrity of general relativity than the linearized approach (and, for this reason, it is a good example of an alternative to the linearized). On the other hand, the canonical approach is, it seems to me, a bit on the simple-minded, naive side. It takes, at its model, elementary Schrödinger quantization of a particle. But quantum theory has advanced considerably since its beginnings. We begin with a brief review of Schrödinger quantization. We then attempt to carry over, as directly as possible, these ideas to general relativity. The result is an imprecise, but suggestive, program for obtaining a quantum theory of gravitation.

Consider a single particle. We can describe the particle by its position x and momentum p. Thus, as the particle moves around in time, the motion is described by $x(t)$ and $p(t)$. The dynamics of the particle are described by a pair of differential equations which express $\dot{x}$ and $\dot{p}$ as functions of x and p. Thus, if one specifies the values of x and p at some initial time, then the equations of motion determine x and p for all future times. The initial data for the particle consist of the values of x and p. It usually turns out that the equations of motion for the particle can be cast into the following form. One can find a certain function $H(x,p)$ of x and p such that

$$\dot{x} = \frac{\partial}{\partial p}H(x,p) \qquad \dot{p} = -\frac{\partial}{\partial x}H(x,p) \qquad (169)$$

If the equations of motion can be cast into the form (169), they are said to be in *Hamiltonian form*. The function $H(p,q)$ is called the *Hamiltonian* of the system.

The Schrödinger quantization scheme is applicable to classical system whose equations of motion have been placed in Hamiltonian form.

179

The initial data, x, p, are replaced by a single, complex-valued wave function, $\psi(x)$. Instead of $x(t), p(t)$, we have $\psi(x, t)$. Thus, the motion of the system in time is described via time-dependence in ψ. The Hamiltonian equations of motion, (169) are replaced by

$$- \frac{\hbar}{i} \frac{\partial}{\partial t} \psi = H\left(x, \frac{\hbar}{i} \frac{\partial}{\partial x}\right) \psi \qquad (170)$$

where $H(x, \frac{\hbar}{i} \frac{\partial}{\partial x})$ means "replace p in $H(x, p)$ by the differential operator $\frac{\hbar}{i} \frac{\partial}{\partial x}$" (a rather vague prescription). Thus, given $\psi(x, t_0)$ for some value of t_0, (170) determines $\psi(x, t)$ for all t.

The Hilbert space for this quantum theory consists of the (complex) vector space of all (sufficiently well-behaved) solutions of (170). One then introduces position, momentum, energy operators, etc.

The idea is, firstly, to try to express the equations of general relativity in "Hamiltonian form". Then, one applies the Schrödinger prescription to obtain a quantum theory. Recall the initial-value formulation of general relativity. The initial data consist of the induced metric h_{ab} and the extrinsic curvature Π^{ab}. These evolve with time according to the equations

$$\begin{aligned}
\dot{h}_{ab} &= 2\varphi \Pi_{ab} \\
\dot{\Pi}^{ab} &= -D^a D^b \varphi - 2\varphi \Pi^{am} \Pi^b{}_m - \varphi \Pi\Pi^{ab} + \varphi \mathscr{R}^{ab}
\end{aligned} \qquad (171)$$

where φ is the evolution function. One first task is to re-express (171) in Hamiltonian form.

Set $p^{ab} = \Pi^{ab} - \Pi h^{ab}$. This p^{ab} is, it turns out, more closely analogous to the p for a particle than Π^{ab}. Rewriting (171) in terms of p^{ab}, we obtain

$$\begin{aligned}
\dot{h}_{ab} &= 2(p_{ab} - \frac{1}{2} p\, h_{ab}) \\
\dot{p}^{ab} &= -D^a D^b \varphi + \varphi \mathscr{R}^{ab} - 2\varphi\, p^a{}_m p^{bm} \\
&\quad + \frac{3}{2} \varphi\, p\, p^{ab} - 2\varphi\, h^{ab} p^{mn} p_{mn} + \frac{1}{4} \varphi\, p^2 h^{ab}
\end{aligned} \qquad (172)$$

We wish to express these equations in Hamiltonian form. This is in fact possible: choose for the Hamiltonian

$$H = - \int_S \varphi\, (\mathscr{R} - p^{mn} p_{mn} + \frac{1}{2} p^2)\, \mathrm{d}V \qquad (173)$$

where the integral extends over the entire 3-manifold S. (We ignore questions of convergence of integrals. We shall also allow ourselves to throw away surface terms at will. Such details are unimportant at this stage of theory-building.) Note that $H(h_{ab}, p^{ab}$ does indeed assign a real number to each choice of data, h_{ab}, p^{ab} , as would want.

Thus, we have a Hamiltonian formulation of the initial-value formulation of general relativity (ignoring for the moment, the question of constraints). Note that we go to the initial-value formulation of general relativity because, in the one-particle discussion, time played a special role. An analogy could be made to general relativity only reintroducing a "time" there. That is precisely what the initial-value formulation accomplishes. In a certain sense we have, already at this stage, violated the spirit of general relativity.

The next step is to write down the wave function. Instead of the $\psi(x)$ in the Schrödinger theory, we have $\psi(h_{ab})$, a complex-valued function of the collection of all positive-definite metrics on the (fixed) three-dimensional manifold S. We wish to permit evolution, so we write $\psi(h_{ab}, t)$. The Schrödinger equation, (170) becomes, using (173),

$$-\frac{\hbar}{i}\frac{\partial}{\partial t}\psi = \left(\frac{\hbar}{i}\right)^2 \left(h^{ac}h^{bd} - \frac{1}{2}h^{ab}h^{cd}\right)\frac{\delta^2\psi}{\delta h_{ab}\,\delta h_{cd}} - \mathscr{R}\,\psi \qquad (174)$$

where, $\delta/\delta h_{ab}$ refers (rather imprecisely) to functional derivatives. Thus, just carrying over the analogy with Schrödinger quantization, we are led to describe the "quantum gravitational field" by a complex-valued function, $\psi(h_{ab}, t)$, on the space of all positive-definite metrics on S, and on t, This function must satisfy the "gravitational Schrödinger equation", (174).

We have, up till now, ignored the constraint equations, (134) and (135). Clearly, one is doing something essentially wrong if he simply ignores certain equations: we have yet fully incorporated Einstein's equation into our theory. First note that, in terms of p^{ab}, the constraints take the form

$$D_b p^{ab} = 0 \qquad (175)$$

$$\mathscr{R} - p^{ab}p_{ab} + \frac{1}{2}p^2 = 0 \qquad (176)$$

The question is: How do we "incorporate" these equations into the theory? The most naive answer is simply to incorporate them as operator equations, using the replacement of p_{ab} by $(\hbar/i)\,\delta/\delta h_{ab}$. Let's try this to see what happens. The classical constrain (175) would then be replaced by the following condition on our wave function

$$D_b \left(\frac{\hbar}{i} \frac{\delta \psi}{\delta h_{ab}} \right) = 0 \qquad (177)$$

To interpret (177) multiply by an arbitrary vector field v_a on S, and integrate by parts to obtain

$$\int\limits_S \left(\frac{\delta \psi}{\delta h_{ab}} \right) (D_{(a} v_{b)}) \, dV = 0 \qquad (178)$$

The validity of (177) is equivalent to the validity of (178) for all v_a. But (178) is easy to interpret. Note that $D_{(a} v_{b)} = \frac{1}{2} \mathscr{L}_v h_{ab}$. Thus, $D_{(a} v_{b)}$ is (up to a factor) the rate of change of h_{ab} under the diffeomorphism generated by motions along the integral curves of v_a. Therefore (by the chain rule), (178) states that the rate of change of $\psi(h_{ab})$, as h_{ab} changes by the diffeomorphism generated by v_a, is zero. To say it another way, (178) requires that, if h_{ab} and h'_{ab} are two metrics on S which differ by a diffeomorphism on S (i.e., if h^{ab} and h'^{ab} are isometric), then $\psi(h_{ab}) = \psi(h'_{ab})$. We can write this symbolically as $\psi = \psi(\text{geometry})$. This conclusion is also reasonable physically. The physics of isometric metrics is identical (the only difference being the labeling of points of S). Thus, one might expect ψ to assume the same value on two such metrics. To summarize, the constraint equation (175) leads to the quantum condition (177) which, geometrically, means that ψ is invariant under replacing h_{ab} by the result of subjecting h_{ab} to a diffeomorphism in S.

We now repeat for (176). It is not hard to guess what the answer will be: (176) requires that $\psi(h_{ab}, t)$ be invariant under motions in time. That this is indeed the case can be seen immediately by noting that the Hamiltonian of our theory, (173), is just an integral of the constraint (176). Hence, the quantum version of (176) is precisely the condition that the right side of (174) vanishes. Thus, we require $(\partial/\partial t)\psi(h_{ab}, t) = 0$, i.e., that ψ in fact be independent of t. (It's a good thing. The interpretation of this "t" was always rather obscure, anyway.)

Thus, the constraints, (175) and (176), are related to the section of diffeomorphisms in space-time. This is expressed, in the quantum theory, by certain invariance of the wave function. In fact, we have seen these notions once before: in the gauge transformations in the linearized theory. These gauge transformations, again represented the action of diffeomorphisms. Thus, gauge (linearized version), action of diffeomorphisms (full theory), constraints (initial-value formulation), and invariance of wave function (quantum theory) all are manifestations of essentially the same thing.

We summarize by stating the formalism of this "theory". The Hilbert space is the space of all complex-valued functions $\psi(h_{ab})$ on the space of positive-definite metrics on S, such that ψ is invariant under the action (on h_{ab}) of diffeomorphisms on S, and such that

$$\left(\frac{\hbar}{i}\right)^2 \left(h^{ac}h^{bd} - h^{ab}h^{cd}\right) \frac{\delta^2\psi}{\delta h_{ab}\,\delta h_{cd}} - \mathscr{R}\,\psi = 0 \qquad (179)$$

We put theory in quotation marks because we have here merely an equation and a few words. What does it all mean? What is the measurement situation? What would it be like to live in such a quantum space-time? What is the correspondence limit? To what extent have the principles of quantum theory been incorporated? To what extent is this theory a "fuzzing out" of general relativity? What alternative formulations are available, and how do they compare with this one?

We are today a good way from satisfactory answers to questions of this sort.

About the author

Robert Geroch is a theoretical physicist and professor at the University of Chicago. He obtained his Ph.D. degree from Princeton University in 1967 under the supervision of John Archibald Wheeler. His main research interests lie in mathematical physics and general relativity.

Geroch's approach to teaching theoretical physics masterfully intertwines the explanations of physical phenomena and the mathematical structures used for their description in such a way that both reinforce each other to facilitate the understanding of even the most abstract and subtle issues. He has been also investing great effort in teaching physics and mathematical physics to non-science students.

Robert Geroch in his office

28219747R00105

Made in the USA
Lexington, KY
10 December 2013